Dry Gardening Secrets for Beginners

Create Vibrant, Eco-Friendly Gardens That Adapt to Climate Change and Boost Well-Being

Richard Dreisdale

To my wife, Ellen H. Moore, whose lifelong
love of gardening inspired this book.

Thanks go out to all those who contributed
their valuable time helping to make this a better book
with their questions, comments, and suggestions.

Contents

Introduction

In the heart of a sun-drenched desert, where rainfall is scarce, a garden thrives. This isn't a mirage but a testament to the resilience and ingenuity of dry gardening. Lush and vibrant, this garden flourishes under techniques that marry beauty with sustainability.

My name is Richard Dreisdale, and this book is born from my journey through two careers: first, as an engineer designing electrical systems, and now, in my golden years, as an advocate for water-efficient gardening. My engineering background instilled a systematic approach to problem-solving, which I now apply to gardening—transforming a personal passion into a mission for environmental sustainability.

Now more than ever, wherever you live in the world, adopting sustainable gardening practices is crucial. With escalating water scarcity and the growing impact of climate change, even areas without previous water stress may face new challenges. This book equips you—whether a homeowner or an apartment dweller—with the knowledge to create a thriving garden that uses minimal water.

I will guide you through efficient irrigation methods, drought-resistant plant selection, and techniques to improve soil retention and health. You'll find practical DIY projects, success stories from gardens in similar climates, and beginner-friendly advice that simplifies dry gardening.

This guide combines my technical expertise with practical gardening knowledge. It includes case studies from rural estates to urban spaces, ensuring strategies tailored to your needs.

You're invited to apply these techniques and join a community of forward-thinking gardeners prioritizing sustainability. As we turn the pages, I'll share the challenges and triumphs of transforming a barren plot into a flourishing oasis. This journey is one of education, inspiration, and community.

Chapter 1

The Foundations of Dry Gardening

As you embark on the path toward creating a sustainable garden in an arid or semi-arid climate, it's vital to understand that dry gardening is about embracing a holistic approach that adapts to your unique environmental conditions and minimizes water use. Whether you find yourself on a sprawling suburban lot or a compact urban balcony, the principles of dry gardening can transform your space into a lush, thriving oasis.

Dry gardening significantly enhances well-being. It encourages physical activity and outdoor exposure, both of which are key for maintaining physical health and reducing stress. Gardening, in general, is an eco-friendly choice that instills a sense of environmental stewardship. Moreover, the success of growing a garden with limited water resources can provide a significant psychological boost, enhancing feelings of accomplishment and connection to nature. Overall, dry gardening not only beautifies arid landscapes but also nurtures the mind and body.

Let's debunk a common myth right from the start: Dry gardening does not confine you to a narrow range of plants or gardening styles. It opens up a world of possibilities where efficiency meets beauty. By understanding and implementing the core principles of water conservation, soil health, and plant selection, you can create a garden that is both resilient and vibrant.

1.1 Defining Dry Gardening: Beyond the Arid Climate

The Broad Scope of Dry Gardening

Dry gardening is actually a systematic approach tailored to reduce water usage while maintaining a garden that thrives in diverse environmental conditions. This method is not just for those living in traditionally arid climates but is a valuable practice for gardeners worldwide who seek to optimize their water usage and adapt to the increasing unpredictability of weather patterns.

Dry gardening involves selecting plants not only for their drought resistance but also for their ability to flourish in the local climate and soil conditions. It integrates water-saving irrigation techniques and soil management strategies that together reduce the need for frequent watering. By focusing on these aspects, dry gardening turns a challenging climate into a thriving garden opportunity.

Addressing Misconceptions

A common misconception is that dry gardening is limited to planting drought-tolerant species like succulents or native shrubs. While these plants can be components of a dry garden, the approach is far more inclusive. You can grow a variety of plants, including some vegetables and herbs, by employing strategic water management and soil enhancement techniques. The key lies in understanding the specific needs of plants and how they interact with the environment, allowing for a broader spectrum of gardening choices.

Core Principles of Dry Gardening

At the heart of dry gardening are three core principles: water conservation, soil health, and appropriate plant selection. Each principle is crucial not only for sustaining garden life but also for creating a landscape that requires minimal intervention. Enhancing soil health through organic amendments and mulching leads to improved water retention and reduced evaporation, making the soil a reservoir that plants can draw moisture from over time. Choosing the right plants for your garden's specific conditions—considering factors like sun exposure, wind patterns, and natural precipitation—further ensures that every drop of water is used efficiently.

Adaptability Across Climates and Settings

One of the most exciting aspects of dry gardening is its adaptability. Whether you are cultivating a rooftop garden in a bustling city or a sprawling landscape in a rural area, the principles of dry gardening can be applied to benefit your specific situation. In urban areas, where heat and limited space complicate traditional gardening, techniques such as vertical gardening, the use of reflective mulches, and the selection of container-appropriate plants can transform balconies and terraces into green retreats. In more expansive rural or suburban gardens, larger-scale implementations like rain gardens or strategically placed windbreaks can significantly enhance the efficiency of water use.

By embracing the principles outlined in this sub-chapter, you are taking the first step towards creating a sustainable and beautiful garden that respects the limitations of your local environment yet bursts with vitality. Whether you're dealing with the baking sun of a desert climate or the unpredictable rain patterns of a temperate zone, dry gardening offers practical solutions that ensure your gardening efforts are both successful and sustainable.

1.2 History and Evolution of Dry Gardening Techniques

Dry gardening, though it may seem like a modern response to contemporary environmental challenges, is deeply rooted in the ancient wisdom of civilizations that thrived in arid landscapes thousands of years ago. These early societies, from the Native Americans in the southwestern United States to the ingenious farmers of the Middle East's Fertile Crescent, developed and refined the first techniques of what we now recognize as dry gardening. They built terraces to prevent soil erosion, used stones to reduce evaporation, and selected plants that were naturally adapted to thrive with minimal moisture. These ancient techniques were not just about survival; they represented a profound understanding of the balance between human needs and the natural world, a harmony that is often sought in today's eco-conscious gardening practices.

As trade and communication expanded the world's interconnectedness, so too did the exchange of agricultural knowledge. Techniques like olla irrigation—using terracotta pots to slowly distribute water—traveled across continents, evolving with each culture's unique conditions and needs. In the arid regions of North Africa, for instance, the use of sunken beds to capture and retain scarce rainfall was developed, a method that has seen a resurgence among modern dry gardeners seeking to maximize water efficiency. These cultural exchanges not only enriched local agricultural practices but also laid the groundwork for a global understanding of sustainable gardening in dry environments.

The integration of modern technology has transformed these ancient practices into highly efficient systems. Today, drip irrigation, a method with its origins in early subsurface clay pipe systems, uses digital timers and regulated drip lines to precisely manage water distribution, minimizing waste and dramatically enhancing plant health. Soil moisture sensors and automated weather stations now assist gardeners in making data-driven decisions about when and how much to water, based on real-time environmental conditions. The fusion of historical wisdom with cutting-edge technology allows for a level of precision and sustainability that ancient gardeners could hardly have imagined.

Looking to the future, the principles of dry gardening are set to play an increasingly critical role as climate change and population growth intensify the pressures on water resources. Innovations in desalination and water recycling technologies promise to expand the possibilities for dry gardening in even the most water-scarce regions. As urban areas continue to grow, vertical and rooftop gardens utilizing dry gardening techniques could become vital components of sustainable urban planning, providing green spaces and local food sources. These prospects highlight not only the adaptability and resilience of dry gardening principles but also their potential to significantly shape our response to global environmental challenges.

By examining the past and looking to the future, it becomes clear that dry gardening is more than a set of techniques; it is a continually evolving dialogue between humans and the earth, a testament to our ability to adapt and thrive in diverse conditions. As we face the challenges of the present and future, the lessons from ancient and modern dry gardening remain a guiding light, showing us that through innovation and respect for nature, we can cultivate not only gardens but also a sustainable future.

1.3 Environmental Benefits: A Global and Local Perspective

The practice of dry gardening, while deeply personal, extends its influence far beyond the confines of one's own garden. It plays a crucial role in addressing some of the most pressing environmental issues of our time, including water scarcity, loss of biodiversity, and the effects of climate change. By adopting dry gardening practices, gardeners in all climates of the world contribute to water conservation, support local ecosystems, build resilience against climate variability, and promote a sustainable lifestyle within their communities.

Water Conservation: A Dual Impact

Water is a finite resource that is becoming increasingly scarce in many parts of the world. Traditional gardening practices often involve high water use, which can strain local water supplies and lead to unsustainable extraction of existing underground water resources. Dry gardening, by contrast, focuses on minimizing water use through efficient irrigation methods and by choosing plants that require less water. For example, a simple change from a conventional sprinkler system to a drip irrigation system can save up to 80% of the water by delivering it directly to the plant's roots, where it's needed most. This not only conserves water but also reduces the gardener's water bill and dependency on municipal water systems.

This conservation has a ripple effect that reaches beyond individual gardens. On a global scale, if more gardeners and landscapers adopt water-efficient practices, the cumulative effect could lead to significant reductions in water usage, contributing to the sustainability of existing water resources. It's a powerful example of how individual actions can align with global sustainability goals in all world climates, but particularly in regions facing severe water shortages.

Ecosystem Support Through Native Planting

The choice of plants in dry gardening goes beyond aesthetics and water conservation. By favoring native plants, dry gardeners create a habitat that supports local wildlife, including pollinators such as bees, butterflies, and birds. Native plants are adapted to the local climate and soil, requiring less water and fewer chemicals to maintain. This not only makes them ideal for dry gardening but also helps maintain the health and balance of local ecosystems.

Consider the monarch butterfly, which relies on native milkweed plants for breeding. By incorporating milkweed into a dry garden, gardeners can provide a critical habitat for these butterflies, supporting their lifecycle and contributing to the biodiversity of the area. This practice helps counteract the loss of native flora and fauna due to urban sprawl and intensive agriculture, which often favor non-native, water-intensive grass lawns and ornamental plants.

Building Climate Resilience

Dry gardens are inherently more resilient to droughts and extreme heat, which are becoming more frequent and severe due to climate change.

These gardens are designed to absorb and retain moisture, reduce runoff, and withstand harsh conditions without requiring extensive human intervention.

This resilience is particularly important as weather patterns become more unpredictable. Gardens that are adaptable to such changes can serve as models for climate-smart agriculture and sustainable landscaping practices. They demonstrate that it is possible to maintain green spaces that not only survive but thrive under challenging environmental conditions, providing cooling green cover that helps reduce the urban heat island effect, a significant problem in many cities around the world.

Personal and Community Impact

Finally, the practice of dry gardening encourages gardeners to think critically about their environmental impact and to take actionable steps toward reducing their carbon footprint. The reduction in water use, combined with the decreased need for chemical fertilizers and pesticides that accompany native and drought-resistant planting, results in a lower overall environmental impact of gardening activities.

On a community level, dry gardening can foster a sense of environmental stewardship and collective responsibility. Community gardens employing dry gardening techniques can serve as educational platforms, demonstrating sustainable practices and encouraging more widespread adoption. They can become centers for community interaction and cooperation, promoting not only ecological benefits but also social cohesion and well-being.

Each garden adjusted to dry gardening principles becomes a testament to what is achievable when individuals take thoughtful, informed action toward living more sustainably. It showcases how, yard by yard, we can contribute to a larger environmental solution, one that supports our planet's ecosystems and ensures a viable, lush world for future generations.

1.4 Understanding Your Garden's Microclimate

One of the most significant concepts encountered in years of gardening is the understanding of microclimates. These unique little climates within your garden can vary dramatically over even small distances. A microclimate can be shaded, sunlit, dry, or damp, and each of these conditions can significantly affect the health and growth of your plants. Recognizing and utilizing these microclimates not only optimizes your garden's water usage and plant health but also turns potential gardening challenges into advantages.

Let's start by identifying these microclimates. Take a walk through your garden at different times of the day and year. Notice where the sun hits directly, where shadows occur, and where the wind sweeps through or where it's blocked by structures or dense foliage. You might find a sun trap, an area that receives prolonged sunlight, beside a south-facing wall, which could be perfect for plants that thrive in constant warmth and light. Conversely, the shaded areas under a large tree can support shade-loving plants that require protection from the harsh midday sun. Wind corridors, often found between buildings or through narrow passages, can dry out the soil quickly; understanding this can guide you to where more robust, wind-tolerant plants should go or where windbreaks might be needed. Each of these observations helps you tailor your garden design to the natural environment, reducing the need for artificial adjustments and interventions.

Soil health is another cornerstone of successful dry gardening. Each type of soil, from sandy to clay, holds water differently, and understanding the type of soil in your garden is crucial. Improving your soil with organic matter such as compost or leaf mold increases its ability to retain moisture and support healthy plant roots. For instance, sandy soils, while providing excellent drainage, often fail to hold moisture and nutrients. By incorporating organic materials, you can enhance its water retention capabilities, making it more hospitable for a broader variety of plants. Regularly check your soil's texture and consider a professional soil test every few years to tailor your soil amendments accurately and maintain its health.

Now, let's talk about water management—crucial in any form of gardening but particularly so in dry gardening. Techniques such as mulching and the use of water-retentive materials like coir or vermiculite can significantly reduce water evaporation from the soil. Mulch, a simple layer of organic material spread over the soil, not only locks in moisture but also suppresses weeds and adds nutrients to the soil as it decomposes. Moreover, appropriate irrigation strategies, such as drip irrigation or the use of ollas, ensure that water is delivered directly to the plant roots where it's most needed, minimizing waste. These methods create an environment where water is conserved and recycling becomes a natural part of the garden ecology.

Finally, choosing and positioning plants according to your garden's microclimates can dramatically increase your garden's efficiency and beauty. Plants that are well-suited to their location will naturally thrive and require less intervention in terms of water, nutrients, and care. For example, lavender and rosemary thrive in hot, sunny spots and need little water once established, making them perfect for those sunny areas in your garden. In contrast, ferns and hostas might be just the right fit for shaded, damp corners. Understanding the specific needs of each plant and matching them to the microclimates in your garden not only ensures plant health but also creates a garden that feels intuitively integrated with the natural landscape.

By tuning into the unique conditions of each part of your garden, you can create a thriving ecosystem that is both beautiful and sustainable. This approach makes your garden a joy to behold and a testament to efficient, environmentally conscious gardening. Remember, each decision you make, from soil preparation to plant selection, shapes not just the aesthetic of your garden but its environmental impact as well.

1.5 Psychology of Dry Gardening: Connecting With Nature

Gardening in general, and particularly in the context of dry environments, extends far beyond the physical acts of planting and watering. It is an immersive experience that touches on the profound intertwining of the gardener's mental well-being with the rhythms of the natural world. Engaging with the soil, nurturing plants from seed to bloom, and adapting to the constraints of water scarcity can transform our understanding of nature and ourselves, offering a unique form of therapy and satisfaction.

The mental health benefits of gardening are well-documented but take on added dimensions in dry gardening. This form of gardening requires patience, foresight, and a deep engagement with the environment. The slow and deliberate pace necessary to cultivate a garden with limited water resources can be meditative, providing a counterbalance to the fast-paced pressures of modern life. Stress reduction is a natural outcome, as the garden becomes a sanctuary of peace and natural beauty, distanced from the noise and clutter of urban existence. The act of caring for plants, watching them grow and thrive under your care, fosters a sense of accomplishment and self-efficacy that can elevate mood and boost overall mental health.

Moreover, dry gardening encourages a deeper connection to the environment. By working with the land and its inherent limitations, gardeners become acutely aware of the natural resources that sustain us. This awareness builds a profound respect and appreciation for the natural world, fostering a relationship characterized by stewardship rather than domination. Gardeners often describe feeling more attuned to the seasons, weather patterns, and the wildlife that interacts with their garden, deepening their connection to the place they inhabit.

This practice also nurtures a sustainability mindset, crucial in today's environmental context. Dry gardening by its nature requires one to consider the long-term impacts of their actions on the ecosystem. The choices made in a dry garden—whether it's selecting native plants, using recycled greywater, or employing soil-enhancing techniques like biochar or vermicomposting—reflect a commitment to sustainability. These choices demonstrate a practical application of environmental responsibility, making the abstract concept of "sustainability" tangible and actionable. Over time, this mindset can extend beyond the garden, influencing other areas of life and encouraging a more thoughtful approach to resource use in general.

Community building through shared values is another significant aspect of dry gardening. Gardening, especially in urban settings, can be a solitary activity, but it doesn't have to be. Community gardens, shared spaces, and even informal networks of neighborhood gardeners provide opportunities to connect with others who share a passion for sustainable gardening. These communities can be incredibly diverse, bringing together individuals from various backgrounds united by their commitment to environmentally friendly gardening practices. In these spaces, gardeners exchange tips, share successes and failures, and support each other's efforts to garden sustainably. This sense of community not only enriches the individual's gardening experience but also strengthens communal bonds and promotes a collective approach to environmental stewardship.

Chapter 2

Planning Your Dry Garden

Imagine stepping into a space where every corner, every plot of soil under your stewardship, resonates with potential. As you stand there, perhaps with the sun warming your face or the cool shade brushing your skin, you're not just seeing an empty space—you're envisioning a future where this very spot thrives under your care, transformed into a vibrant testament to dry gardening. This chapter is about turning that vision into a reality, beginning with understanding and optimizing the unique characteristics of your garden space.

2.1 Analyzing Your Space: Sun, Shade, and Soil

Site Assessment

The first step in planning your dry garden is to conduct a thorough site assessment. This process involves observing and documenting various aspects of your garden area, focusing on sunlight exposure, shaded regions, and soil type. Start by tracking how sunlight moves across your space throughout the day. Which areas receive full sun, and which are shaded? Sunlight is a key factor in plant health and water usage; plants that are well-suited to their lighting conditions require less intervention and maintenance.

Next, consider the shaded areas in your garden. Shade can be a valuable asset in a dry garden, reducing the evaporation of precious water from both soil and plant surfaces. However, too much shade might limit your plant choices or require adjustments to your garden layout. Understanding these patterns of light and shade will not only influence your plant selection but also your irrigation planning, ensuring that water is used most efficiently where the sun is strongest.

Finally, assess your soil type. Is it sandy and quick to drain, or is it clay-rich and slow to release water? Perhaps it's a loamy balance between the two? Take a handful of moist soil and squeeze it in your palm. If it forms a loose ball that crumbles easily, you have sandy soil. If it forms a tight, hard ball, your soil is predominantly clay. Loamy soil, ideal for most garden plants, will form a soft ball that breaks apart gently. Understanding your soil type is crucial because it directly affects water retention and nutrient availability, which are pivotal in a dry garden.

Maximizing Potential

Whether your garden space is a generous backyard or a modest balcony, each offers unique opportunities for dry gardening. For larger areas, consider creating zones based on sunlight exposure and soil type, which can help in planning both planting and irrigation more effectively. In smaller spaces or balconies, vertical gardening can be a superb solution, allowing plants to grow upwards and make the most of your limited square footage. Utilizing hanging baskets, wall-mounted planters, or trellises for climbing plants are all ways to enhance your growing area without expanding horizontally.

Soil Improvement

Improving your soil's health and structure is next. If you have sandy soil, incorporate organic matter like compost or well-rotted manure to increase its water-holding capacity. For clay soils, adding coarse sand and organic matter can improve drainage and prevent water logging. Regularly adding organic matter is beneficial for all soil types, enhancing microbial life within the soil and slowly releasing nutrients as it decomposes. This practice not only feeds your plants but also improves soil structure to better hold water and air, crucial elements for root health.

Design Considerations

When designing your dry garden, aesthetics go hand-in-hand with functionality. Start by sketching a simple map of your garden, marking areas of different light exposure, and noting the soil conditions. This map will serve as a blueprint for your garden design, helping you visualize where to place certain plants and where to install irrigation systems, such as drip lines or ollas. Consider the views from inside your home and from the garden itself. Strategic placement of plants can enhance not just the garden's beauty but also your enjoyment of the space. Integrate paths or stepping stones to allow easy access for maintenance without compacting the soil. Choose materials that reflect the style you wish to achieve but also consider their functionality in a dry environment. For example, lighter-colored materials can reflect sunlight and help keep the area cooler.

Visual Element: Interactive Garden Mapping Tool

To aid in your garden planning, I recommend using an interactive garden mapping tool. This digital tool allows you to input the dimensions of your garden, log varying conditions, and virtually place plants and features to see how they fit together. It's a dynamic way to experiment with different layouts without physical effort, and it can be adjusted as your garden grows and changes over the seasons.

By taking the time to thoroughly understand and thoughtfully plan your space, you lay the groundwork for a garden that is not only aesthetically pleasing but also deeply water-efficient. It's about making every choice count, from the position of each plant to the path of each drop of water. Remember, a well-planned garden is the first step to a successful dry garden, where beauty and sustainability are in perfect harmony.

2.2 Designing Your Dry Garden Layout

When you begin to design your dry garden, think of it as painting a landscape where every stroke serves a purpose, not just for beauty, but for sustainability. The concept of zoning in garden design is crucial, especially in dry gardening where water conservation is key. Zoning refers to organizing your garden space such that plants with similar water needs are grouped together. This strategy not only simplifies watering chores but also ensures that each plant receives just the right amount of moisture without waste. For instance, high water-use plants can be planted near a water source or where your greywater outlets are, while drought-tolerant plants can occupy the furthest, sunniest corners.

This approach minimizes water loss and maximizes plant health, as each zone can be watered according to its specific needs, and you can adjust your irrigation methods accordingly. Drip irrigation might be ideal for one zone, while another might only require occasional hand watering. This tailored watering not only saves water but also prevents the stress that over- or under-watering can cause to plants, promoting a healthier garden overall. Imagine zones as different rooms in a house, each designed to create the perfect environment for its occupants. By zoning your garden, you're essentially building an infrastructure that supports each plant's growth and sustainability.

Balancing functionality with aesthetic appeal is another cornerstone of dry garden design. It's about creating a space that's both beautiful and practical. While the functionality of a dry garden focuses on water conservation and plant health, its aesthetic appeal can be achieved through the thoughtful placement of color, form, and texture. Use drought-resistant plants not only for their resilience but also for their beauty—many drought-tolerant plants offer vibrant flowers and unique foliage.

Arrange plants in patterns or clusters that draw the eye, using color to create depth and interest. For example, a backdrop of green succulents against which bright, drought-tolerant flowers bloom can create a visually striking contrast.

Incorporating hardscaping elements like paths, rocks, and mulch not only adds visual interest but also reduces water use. Paths provide structure and define different areas of your garden, guiding visitors through the space and discouraging foot traffic in growing areas, which can compact soil and reduce its water absorption. Rocks and pebbles add texture and color, and when used as mulch, they help retain soil moisture and suppress weed growth. Mulch, whether organic like bark and straw or inorganic like gravel, plays a dual role—it conserves water by reducing evaporation and adds a polished look to the garden beds.

Creating a wildlife-friendly garden not only enhances the beauty and diversity of your garden but also promotes ecological balance. Features that attract and support local wildlife, such as birds and pollinators, include native flowering plants, shallow water baths, and insect hotels. These features can turn your garden into a lively habitat and help with the pollination of your plants, enhancing your garden's productivity and resilience. Planting native wildflowers and shrubs not only provides food and shelter for local wildlife but also brings color and movement into the garden as butterflies, bees, and birds become regular visitors. Consider incorporating a small water feature, like a birdbath or a small pond, which can serve as a focal point and a vital resource for wildlife, especially in arid areas where water is scarce.

By thoughtfully designing your dry garden to include these elements, you create a space that is not only a model of water efficiency but also a place of beauty and refuge for both you and the local wildlife. Each design choice should be a step towards a more sustainable, vibrant garden that thrives with minimal water input, aligning with the broader goals of conservation and ecological mindfulness. Such a garden not only conserves precious resources but also becomes a sanctuary for both the gardener and the native wildlife, a place where nature and nurture meet.

2.3 Choosing the Right Plants for Your Climate

When the time comes to pick plants for your dry garden, the selection isn't just about what catches your eye—it's about what will thrive in your garden's unique conditions. The smart choice involves opting for native plants that are well-adapted to the local environment and require less water. Native plants have evolved over thousands of years to adapt to the specific climate, soil, and even the local pests of their natural habitat. This long-term adaptation makes them a perfect fit for sustainable landscaping, as they require far less water, fertilizer, and pest control than their non-native counterparts. For example, if you're gardening in the Southwestern United States, plants like the blue palo verde (*Parkinsonia florida*) or the desert marigold (*Baileya multiradiata*) can not only survive but thrive, with minimal intervention.

The choice of drought-tolerant plants extends beyond natives, encompassing a plethora of species that are known for their ability to survive in dry conditions. Succulents like Sedum or Sempervivum are renowned for their water-storing leaves and can be a visually striking addition to any dry garden. Other drought-resistant favorites include lavender, Russian sage, and the blanket flower, which all bring color and variety while requiring minimal moisture. The key here is to understand not just the water requirements but also the sunlight and soil needs of these plants to ensure they can perform at their best in your garden.

Matching plants to specific conditions is crucial for minimizing water usage and maximizing plant health. This matching goes beyond simply choosing drought-tolerant species; it involves a detailed understanding of your garden's microclimates, which you've previously mapped. For instance, a plant that thrives in full sun and requires well-draining soil should be placed in a part of your garden that meets these conditions. This strategic placement reduces the need for supplemental watering and increases the plant's chances of thriving. Additionally, consider the mature size of the plants to avoid overcrowding, which can lead to competition for water and nutrients.

Many people who have moved to an arid region from a non-arid one would like to include a few varieties from their former homes in their new dry garden. To determine which plants from your previous home will thrive in your new arid location, start by looking at their water needs. Plants from wetter areas usually require more moisture than is available in an arid climate, so it's important to find out if they are drought-tolerant. Also, check how much sun and heat they can handle, as arid regions tend to have intense sunlight and high temperatures. If the plant needs a lot of water or struggles with heat, it may not do well in your new environment.

There also may be microclimates in various locations in your new dry garden that approximate conditions in your former area closely enough to allow your old favorites to survive.

Additionally, consider the type of soil and how fast it drains. Arid regions often have sandy or rocky soil, which doesn't hold water as well as the loamy, nutrient-rich soils found in many non-arid areas. Plants that prefer rich, moist soil might struggle unless you make adjustments, such as using mulch to conserve moisture. You can also look for plants similar to those from your old home that are known to thrive in hot, dry climates.

The concept of layering and fostering diversity in your garden plays a significant role in creating a resilient and water-efficient space. Layering in gardening refers to planning your garden in such a way that plants of different heights, growth rates, and rooting depths coexist. This practice not only creates a visually appealing, multi-dimensional space but also promotes a healthier garden ecosystem. For example, deep-rooted plants can help bring nutrients up from the lower soil layers, which benefits the shallow-rooted plants around them. Moreover, a diverse garden with a mix of different types of plants is less likely to succumb to pest infestations or diseases, as the variety itself helps to prevent the spread of pests and pathogens.

Think of your dry garden as a tapestry, with each plant adding its unique thread to the overall design. Each choice you make—from the native species that require less water, to the exotic succulents that store moisture in their leaves—contributes to a garden that is not only a pleasure to behold but also a model of environmental sustainability. As you select each plant, consider its role not just in your garden's aesthetics but also in its ecosystem.

2.4 Creating a Water Budget for Your Garden

Managing water usage effectively starts with a clear understanding of your garden's unique water needs. This understanding hinges not just on the types of plants you've chosen, but also on the specific climate conditions of your area. To begin estimating a water budget, calculate the evapotranspiration rate—this is a measure of how much water your plants lose to the atmosphere and how much they use for growth. This rate varies widely based on factors such as temperature, humidity, wind, and the type of vegetation. Local agricultural extensions or online gardening resources often provide evapotranspiration rates for specific regions, which can be a great starting point.

To tailor this to your garden, list all your plants and categorize them by their water needs—low, medium, or high. Consider the microclimates within your garden that you've previously identified, as these will affect water requirements as well. For instance, plants in full sun and windy conditions may need more frequent watering compared to those in shaded or protected areas. Summing the needs of all these plants gives you a baseline for your garden's total water requirement. However, remember that nature is dynamic. Weather changes, plant growth, and other factors will require you to adjust this baseline periodically, making flexibility a key component of water budgeting.

Efficient watering practices are crucial in maximizing the effectiveness of every drop of water you use. Timing, for instance, plays a pivotal role. Watering in the early morning or late evening reduces water loss to evaporation. Using soaker hoses or drip irrigation systems can deliver water directly to the root zones where it's most needed, minimizing wastage through runoff or evaporation. Another strategy is to use a cycle-and-soak method, especially if you have sloped areas; this involves multiple short watering cycles to allow water to soak in thoroughly, preventing runoff and ensuring deeper water penetration into the soil.

Rainwater harvesting and rain gardens can provide beautiful and water efficient landscape features. See more details in Chapters 3 & 7.

Monitoring and adjusting your garden's water usage is the final, ongoing step in establishing a water-efficient garden. Regularly check soil moisture using either a simple manual soil probe, or install soil moisture sensors that can provide real-time data. Adjust your irrigation schedules based on these readings and on changing weather conditions. For instance, after a rainstorm, you might skip your next scheduled watering, or during a heatwave, you might increase the water volume slightly. Keeping a garden journal can be helpful here, allowing you to record watering times, weather conditions, and plant responses over time, helping you fine-tune your watering schedule to your garden's specific needs.

Through these methods, you ensure that your garden thrives and contributes to broader environmental conservation efforts. A well-planned water budget supports your garden's health and also aligns with sustainable practices that benefit the larger community. As you adjust to seasonal changes and the evolving needs of your garden, this dynamic approach to water management becomes a rewarding part of your gardening experience.

2.5 Incorporating Permaculture Principles

Permaculture is a design philosophy and approach to land management that adopts arrangements observed in flourishing natural ecosystems. It aims to create sustainable and self-sufficient agricultural systems by integrating the land, resources, people, and the environment through mutually beneficial synergies. The term, a contraction of "permanent agriculture," was coined by Australians Bill Mollison and David Holmgren in the 1970s.

Permaculture emphasizes the harmonious interplay between the natural world and human-made systems, allowing them to function cohesively to support one another. It involves designing ecological landscapes that produce food, reduce waste, and conserve resources.

Key principles of permaculture include:

1. Observation and interaction: Understanding natural processes and systems through close observation before making any interventions.

2. Energy conservation: Making the most efficient use of resources by designing systems to minimize waste and energy consumption.

3. Natural resource cycles: Ensuring that resources (such as water, nutrients, and organic matter) are used effectively and continuously cycled back into the system.

4. Supporting biodiversity: Promoting a variety of species to ensure a resilient and adaptive ecological system in the garden.

5. Integration rather than segregation: Designing systems that interconnect various elements to support one another for optimal functioning.

Permaculture can be used not only in agriculture but also in urban settings, where it influences building design, community living, and urban planning to create sustainable and self-supporting habitats.

As we wrap up this chapter on planning your dry garden, we've explored a comprehensive approach that covers everything from the initial analysis of your garden space to the thoughtful integration of permaculture principles. Each section has built upon the last, culminating in a holistic view of garden planning that emphasizes sustainability, efficiency, and community engagement. This foundation sets the stage for the next steps in your gardening adventure, where you will delve into the specifics of implementing your garden design, managing your water resources wisely, and ensuring your garden thrives.

Chapter 3

Water-Saving Irrigation Strategies

Imagine if every drop of water you used in your garden could tell a story. What would it say? Would it speak of a journey where it quenches the thirst of a budding tomato plant or tells a tale of waste as it evaporates from the surface of an overwatered lawn? In this chapter, we turn our focus to the lifeblood of any garden—water. Specifically, how to manage it with precision and care, ensuring that every drop not only counts but also contributes to the thriving ecosystem you are cultivating in your dry garden.

3.1 Water Efficiency of Drip Irrigation

Drip irrigation is an efficient watering system widely used in gardening that delivers water directly to the roots of plants through a network of valves, pipes, tubing, and emitters. This method minimizes water wastage and enhances plant growth by ensuring that water is slowly and precisely applied to the soil. It is particularly advantageous in arid regions or in situations where water conservation is crucial. By maintaining optimal moisture levels in the soil, drip irrigation not only conserves water but also reduces the proliferation of weeds and mitigates the risk of plant diseases associated with overwatering. This targeted approach to irrigation is instrumental in promoting sustainable plant health. See Chapter 7 for a couple of economical examples of DIY Drip Irrigation Systems.

3.2 Using Ollas (clay pots) for Efficient Watering

The olla (oh-yah), an unassuming clay pot, is hardly a new invention, yet it offers timeless benefits, especially in the realm of dry gardening. These traditional irrigation tools have been used for centuries, primarily in arid regions across the globe, to provide a steady, controlled supply of water directly to the roots of plants. The principle behind their use is elegantly simple: the porous nature of unglazed clay allows water to seep slowly into the soil, thereby providing moisture directly to plant roots with minimal evaporation or waste. This method of irrigation is not only efficient, but also based in the sustainable use of resources, making it a perfect fit for modern gardens that aim to minimize water use while ensuring plant health.

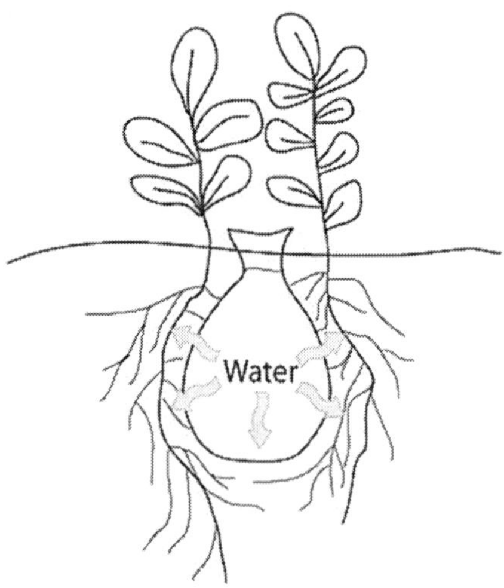

nativeseeds.org/blogs/blog-news/how-to-use-olla-irrigation#

2024 Native Seeds Search

The effectiveness of ollas in dry gardening lies in their ability to maintain a consistent moisture level around the roots, which is crucial for optimal plant growth. Traditional watering methods, such as overhead sprinkling, often lead to uneven moisture distribution, with some areas receiving too much water and others not enough. Moreover, much of the water can evaporate before it even reaches the plant roots. Ollas mitigate these issues by slowly releasing water into the surrounding soil at a rate that matches the absorption capability of the soil and the needs of the plants. This targeted watering ensures that plants receive exactly what they need, when they need it, leading to healthier growth and less water waste.

For those interested in incorporating ollas into their garden, the setup process is straightforward. While you can purchase ready-made ollas, many gardeners enjoy the process of creating their own from recycled or locally sourced clay pots. See Chapter 7 for information on using ollas in your garden.

The use of ollas embodies the principles of sustainability and efficiency that are central to dry gardening. By adopting this ancient yet effective technology, you are participating in a time-honored tradition of resource conservation, ensuring that every drop of water serves a purpose in nurturing the life within your garden. As you integrate ollas into your gardening practice, you'll not only see the benefits in the health and vigor of your plants but also in the resilience of your garden as a whole.

3.3 Rainwater Harvesting and Utilization in the Dry Garden

Rainwater harvesting is akin to capturing a bit of the sky—a resource often overlooked but immensely valuable, especially in regions where water is scarce. Imagine the satisfaction of nurturing your garden using water you've collected, a resource that's both free and environmentally friendly. The essence of rainwater harvesting lies in its simplicity and the profound impact it can have on your garden's health and your water bill. Ancient civilizations in the Middle East and other arid regions owed their very survival to the collection of the rainwater that fell so infrequently.

The environmental impact of using rainwater over tap water for garden irrigation extends beyond mere water conservation. First, rainwater is free from the salts and chemicals typically found in municipal water supplies, which can accumulate in the soil and potentially harm plant roots over time. Using rainwater helps maintain the natural soil balance, promoting healthier plant growth and reducing the need for chemical fertilizers. Moreover, by reducing your reliance on treated water, you lessen the energy and chemical use associated with water treatment and distribution—a significant environmental benefit. Furthermore, rainwater harvesting mitigates stormwater runoff, a major cause of erosion and water pollution in urban areas. By capturing stormwater and using it in your garden, you help reduce the runoff that carries pollutants into local waterways, thus protecting aquatic ecosystems and improving water quality.

The regulatory landscape for rainwater harvesting varies significantly from one region to another. In some areas, stringent regulations govern the use and treatment of harvested rainwater. Familiarizing yourself with these regulations and ensuring that your system complies with local laws is essential for a smooth and lawful operation. Consulting with professionals during the planning and installation phases can provide valuable insights into the best practices and legal requirements specific to your area, helping you design a system that is not only efficient but also compliant.

Incorporating rainwater harvesting into your garden practice is not just about sustainability. It's a proactive step towards a more sustainable lifestyle. As you watch the rainwater slowly drip through your garden's ecosystem, you'll appreciate not just the beauty and productivity of your garden but also the cycle of nature that sustains it. See Chapter 7 for hints on how to incorporate a rain harvesting setup for your home garden.

3.4 Utilizing Greywater Without Compromising Plant Health

In our quest to maximize every drop of water in our gardens, let's turn our attention to an often-overlooked resource: greywater. This is water that has been gently used in your household, coming from sources like sinks, showers, and laundry machines. It's not suitable for drinking, but with proper management, it can be a gold mine for garden irrigation, especially in dry climates where every ounce of water counts.

Understanding greywater is the first step toward integrating it into your garden's watering regimen. Essentially, greywater includes all domestic wastewater except that from toilets, which is known as blackwater. Depending on your household habits, greywater can make up a significant portion of your overall garden water usage, and reusing it not only reduces your water bills but also lessens the strain on municipal water systems and local watersheds. However, the key to successful greywater use lies in ensuring that it is safe for your plants and soil. This means being mindful of the products you use daily. For instance, biodegradable, low-sodium soaps and detergents are crucial because standard cleaning products often contain salts and chemicals that can build up in the soil over time, potentially harming plant roots and microbial health.

To safely redirect greywater to your garden, start by setting up a simple diversion system. This could be as straightforward as attaching a hose to the outlet pipe of your washing machine. This hose then directs the water into a mulch basin or directly to water-tolerant plants in your garden. Mulch basins are particularly effective as they filter the water naturally through layers of wood chips or straw, which trap any particulates and break down potential pollutants before the water reaches the plant roots. It's important to design your system so that greywater can be switched back to the sewer or septic system when necessary, such as when washing heavily soiled items that might contain contaminants unsuitable for garden use.

Here is a simple way of collecting greywater manually: Collect rinse water from kitchen and bathroom sinks, collect cooking water, vegetable rinse water, and left over beverages from the table. These can all be collected in containers and combined, then taken out to the garden periodically. Do not include meat scraps, meat juices, or gravies.

Long-term management of greywater involves regular monitoring of soil health to prevent the accumulation of salts and chemicals. This is particularly vital in dry gardening, where soil conditions directly influence water efficiency and plant health. Periodic soil testing can be beneficial to track changes in soil composition, allowing you to adjust your greywater usage and treatment accordingly. Adding organic matter regularly to your soil can also help; compost not only enriches the soil but also improves its structure and ability to handle the regular influx of greywater without becoming waterlogged or chemically imbalanced.

Adopting greywater usage in your garden might seem like a small step, but it represents a significant stride towards a more sustainable and self-sufficient gardening practice. By reusing water that would otherwise go to waste, you're contributing to a larger ethos of resource conservation, crucial in today's environmental climate. As you continue to explore and refine your greywater system, you'll not only see the benefits in your flourishing garden but also feel a deeper connection to the sustainable practices that support it. Chapter 7 explains the details of setting up your own greywater system.

3.5 Mulching Techniques to Retain Soil Moisture

Mulch is more than just a decorative layer in your garden; it's a crucial component of your water conservation strategy, especially in arid climates. When applied correctly, mulch significantly enhances soil moisture retention, decreases evaporation rates, and helps suppress weed growth that competes with your plants for precious water resources. This isn't just about keeping your soil damp longer; it's about creating an environment where water is conserved and managed wisely, ensuring that your plants remain hydrated and healthy with minimal additional input.

The benefits of mulching are manifold. By covering the soil, mulch acts as a barrier between the sun and the soil surface, reducing water loss due to evaporation. This is particularly vital during the peak of summer when temperatures soar, and the sun is relentless. Additionally, mulch helps maintain an even soil temperature, protecting roots from extreme heat and cold, which can stress plants and make them more susceptible to disease. Mulch discourages weed germination and growth by blocking sunlight, limiting space, and preserving moisture to be absorbed by your garden plants.

Choosing the right type of mulch for your garden depends on several factors, including the specific needs of your plants, the climate, and your personal aesthetic preferences. Organic mulches, such as straw, bark chips, shredded leaves, or grass clippings, are beneficial because they decompose over time, adding organic matter and nutrients back into the soil. This process also encourages the activity of beneficial soil organisms, such as earthworms, which improve soil structure and nutrient availability. Inorganic mulches, like gravel, stones, or landscape fabric, don't enrich the soil but can be more permanent solutions that require less frequent replenishment. They are particularly effective in pathways or around plants that prefer a drier soil environment.

When applying mulch, there are several best practices to consider to maximize its effectiveness. First, ensure that the soil is moist before you apply mulch. Applying mulch over dry soil can actually prevent rain or irrigation water from penetrating the soil surface, which can lead to drier conditions than you started with. Apply a layer of mulch about two to four inches thick, which is enough to provide the benefits of moisture retention and weed suppression without suffocating your plants. Be careful to leave a small space around the base of each plant; this prevents moisture buildup around the stems, which can lead to rot or fungal diseases. Also, consider the aesthetics of your garden when applying mulch. A uniform layer not only looks tidy but also ensures even coverage and effectiveness.

Seasonal changes affect how mulch works in your garden, and adjusting your mulching strategy accordingly can lead to better water conservation and plant health. In spring, delay mulching until the soil has warmed up. Applying mulch too early can keep the soil cold and slow down plant growth. In summer, replenish mulch as needed to ensure it maintains its depth and effectiveness during the hottest months. In autumn, consider adding a fresh layer of mulch to protect plant roots from freezing temperatures and to provide nutrients for the following growing season. In winter, check that the mulch hasn't compacted or become matted down; fluff it up if necessary to ensure it continues to protect the soil from winter cold and moisture loss.

By integrating these mulching techniques into your garden care routine, you not only enhance your garden's beauty and productivity but also contribute to a larger effort of water conservation and sustainability. As you layer your garden with mulch, think of it as insulating a cherished home: you're not just decorating the exterior, you're protecting and nurturing the life within, ensuring that it thrives through the seasons in a harsh climate.

As we wrap up this exploration of mulching techniques, remember that these strategies are part of a broader commitment to sustainability in your dry garden. By understanding and implementing effective mulching practices, you're taking significant steps towards creating a garden that is not only lush and productive but also environmentally responsible. This chapter sets the stage for further discussions on how to select and care for drought-resistant plants, ensuring that your garden remains a beacon of sustainability in an increasingly dry world.

Chapter 4

Selecting and Caring for Drought-Resistant Plants

Imagine stepping into a garden where each plant not only survives but thrives, despite the blazing sun and sparse rainfall. This isn't the result of constant care or daily watering but the strategy of choosing the right plants—those that are built to withstand dry conditions. In this chapter, we delve into the world of drought-resistant plants, exploring how their unique adaptations can be the cornerstone of a flourishing garden in arid climates. From deep-rooted shrubs to succulent beauties, these plants bring resilience and beauty to your garden, reducing your water usage without sacrificing the lush, vibrant landscape you desire.

4.1 Top Drought-Resistant Plants for Your Garden

Diverse Plant Selection

In the realm of drought-resistant gardening, diversity is not just a visual benefit but a functional strategy. Integrating a variety of drought-resistant plants into your garden design ensures that you have a robust, resilient landscape capable of adapting to varying weather conditions. These plants range from hardy perennials like lavender and Russian sage to versatile succulents such as aloes and sedums. Each plant has evolved to survive in environments where water is a luxury, making them perfect candidates for your water-wise garden.

Consider the vibrant echinacea, or coneflower, known for its striking, daisy-like appearance and ability to withstand drought once established. Or the agave, with its dramatic foliage, acting not only as a focal point in your garden but also as a model of ultra-low-water gardening. The key is to select plants that not only tolerate dry conditions but also complement each other in form, color, and texture, creating a garden that is both aesthetically pleasing and environmentally sustainable.

Adaptation and Resilience

The true strength of drought-resistant plants lies in their remarkable adaptations. These plants often feature deep root systems that tap into underground water sources far beneath the soil surface, making them less reliant on surface watering. Their leaves may be waxy or hairy, adaptations that reduce water loss by slowing evaporation. Some, like the stonecrop, store water in their thick, fleshy leaves, allowing them to endure prolonged dry spells.

Take, for instance, the yucca, which exemplifies resilience with its sword-like leaves that channel dew and rainfall straight to its root base. Or the cacti, perhaps the champions of drought resistance, whose thick, succulent walls store copious amounts of water. Each of these plants not only survives but thrives in conditions that would stress typical garden varieties, making them invaluable in your dry garden.

Aesthetic Considerations

Integrating drought-resistant plants into your garden goes beyond their water-saving benefits; it's also about creating a visually captivating space. These plants offer a range of textures, colors, and forms that can enhance the aesthetic of any garden. The silver-gray foliage of lamb's ears or the vivid blooms of the pincushion flower add splashes of color and contrast, providing visual interest throughout the growing season.

Arranging these plants in your garden should be a creative exercise, akin to painting a landscape. Group plants with similar hues for a harmonious effect or mix contrasting colors and textures to create a dynamic composition. Remember, the goal is to balance beauty with sustainability, creating a garden that delights the senses while conserving resources.

Plant Profiles: Key Drought-Resistant Varieties

Here's a categorized list of drought-resistant plants based on their water needs and preferred microclimate:

Low Water Needs

These plants thrive with minimal water once established and prefer full sun conditions.

1. **Lavender (Lavandula):** Thrives in full sun and well-drained soil. Prefers hot, dry climates. Once established, it needs minimal watering and adds fragrant, attractive blooms to the garden.

2. **Russian Sage:** Needs full sun and can tolerate very dry conditions. Its tall spikes of purple flowers are excellent for attracting pollinators.

3. **Yucca:** Adapts to desert and arid conditions, requiring full sun and little water.

4. **Agave:** Prefers full sun and well-drained soils. It's known for its large rosettes of tough, evergreen leaves.

5. **Cacti:** Best suited for very dry, sunny areas. They require very little water and thrive in arid conditions.

6. **Blanket Flower (Gaillardia):** Thrives in full sun and tolerates poor, sandy soils. Prefers low to moderate watering once established, with bright, long-lasting flowers.

Moderate Water Needs

These plants can manage with occasional watering and some humidity, thriving best with regular but not frequent irrigation.

1. **Aloe:** Prefers bright, sunny spots and needs occasional watering, more in summer and less in winter.
2. **Echinacea (Coneflower):** Thrives in full sun to partial shade; requires some watering to establish, then becomes more drought tolerant.
3. **Lamb's Ears:** Best in full sun to partial shade; requires well-drained soil and moderate watering.
4. **Pincushion Flower:** Needs full sun and moderate watering, especially during the blooming season to prolong flowering.
5. **Yarrow (Achillea):** A tough perennial for full sun and poor to average, well-drained soils. Drought-tolerant once established, requiring moderate watering with feathery foliage and clusters of small flowers.

Variable Water Needs Depending on Variety

These plants have species that vary in their water needs, generally preferring low to moderate watering once established.

1. **California Poppy:** Does well in full sun and can tolerate poor, dry soils but performs best with moderate watering.
2. **Stonecrop (Type of Sedum):** While generally drought-tolerant, some varieties prefer slightly more frequent watering to maintain their lush appearance. These succulents are excellent for rock gardens.
3. **Desert Marigold (Baileya multiradiata):** A perennial with bright yellow, daisy-like flowers and gray-green foliage.

It thrives in full sun and well-drained soil, providing vibrant color to arid gardens.

4. **Red Yucca (Hesperaloe parviflora):** Red Yucca produces long, arching flower spikes with tubular red or pink flowers. Its grass-like leaves are evergreen and provide year-round interest. It is highly drought-tolerant and prefers full sun.
5. **Blue Fescue (Festuca glauca):** An ornamental grass with fine, blue-gray foliage forming neat, compact mounds. It prefers full sun and well-drained soil and adds texture and color to drought-resistant gardens.

Each plant on this list is highly resilient and suited for gardens that face water scarcity, though their individual needs for sun, soil type, and minimal water requirements vary, enabling them to adapt to various garden environments.

Textual Element: Reflection Section

Reflect on the unique conditions of your garden. Which areas receive the most sunlight? Where is the soil driest? Use this information and that collected in your garden audit in Chapter 2 to think about how you can incorporate drought-resistant plants into your landscape. Consider not only the practical aspects, such as water conservation and soil type but also the aesthetic—how these plants can enhance the beauty of your garden. Reflecting on these questions will help you make informed decisions that align with both your environmental goals and your design aspirations.

As you explore the world of drought-resistant plants, remember that each choice you make contributes to the sustainability and beauty of your garden. These plants are not only survivors; they are thrivers, offering a palette of possibilities that can transform your garden into a resilient, water-wise, and stunning landscape.

4.2 Native Plants: Low-Maintenance and Eco-Friendly Choices

Embracing native plants in your garden is like inviting old friends to a homecoming party. These plants, inherently attuned to the local climate and soil, bring along numerous benefits that extend beyond their beauty. They are naturally engineered to thrive in your garden's specific conditions, which translates into lower water needs and less maintenance. Additionally, they play a crucial role in supporting and sustaining local wildlife and biodiversity.

But perhaps one of the most compelling reasons to integrate native plants into your garden is their role in supporting local wildlife. These plants are a critical part of the local ecosystem, providing nectar, pollen, and seeds that feed everything from bees and butterflies to birds and mammals. Moreover, they offer shelter and breeding grounds for various species. For instance, the milkweed plant is the only host plant for monarch butterfly larvae, while native grasses can provide habitat for ground-nesting birds. By choosing native plants, your garden can become a sanctuary for these creatures, many of which have suffered from habitat loss due to urbanization and agriculture.

Finding the right native plants for your garden can be a delightful exploration of your region's natural heritage. Start by visiting local nurseries, particularly those that specialize in native species. These nurseries often have knowledgeable staff who can offer advice on choosing plants that will suit your specific garden conditions. Local botanical gardens and extension services can also be invaluable resources, offering plant sales, workshops, and guides that can help you make informed choices. Additionally, joining local gardening clubs or societies can provide insights and access to native plants that are known to perform well in your area.

Once you have selected your native plants, integrating them into your garden should be done with consideration of their natural habitats. Observe where these plants grow in the wild—are they basking in full sun, nestled in shade, or thriving in wet areas? Mimicking these conditions in your garden will help ensure their success. For instance, if you're planting a native shrub that thrives under the canopy of tall trees, position it similarly in your garden, perhaps under a larger tree or in a spot that receives partial shade. When planting, consider the mature size of the plants to avoid overcrowding and to reduce competition for water and nutrients.

Incorporating native plants into your garden offers a seamless blend of beauty, ease, and ecological responsibility. It allows you to create a vibrant, healthy garden that demands less from you and gives back more to the environment. As you watch your garden mature, you'll not only enjoy the array of colors and textures that native plants bring but also the buzz of insects, the flutter of butterflies, and the songs of birds that come to call your garden home—a living testament to the beauty and resilience of nature's designs.

Plant Profiles: Key Native Plant Varieties

Desert plants have adapted in fascinating ways to survive in harsh, arid environments. They can be categorized based on their water needs and the specific microclimates they thrive in. Here's a list of native desert plants, grouped accordingly:

Low Water Needs & Hot, Sunny Microclimates

1. **Saguaro Cactus (*Carnegiea gigantea*):** Iconic cactus of the Sonoran Desert, towering and slow-growing.

2. **Golden Barrel Cactus (*Echinocactus grusonii*):** Globular cactus with prominent ribs and bright yellow spines.
3. **Joshua Tree (*Yucca brevifolia*):** A distinctive tree-like plant native to the Mojave Desert with spiky leaves.

Moderate Water Needs & Partial Shade Microclimates

1. **Ocotillo (*Fouquieria splendens*):** Known for its spiny, cane-like branches that burst into bright red flowers.
2. **Texas Sage (*Leucophyllum frutescens*):** Bushy shrub with silvery foliage and purple flowers, thriving with slightly more water.
3. **Chuparosa (*Justicia californica*):** Produces vibrant red flowers that attract hummingbirds. Prefers some shade.

Low Water Needs & Cooler, Elevated Microclimates

1. **Pinyon Pine (*Pinus edulis*):** A small, slow-growing pine tree that thrives in higher desert elevations.
2. **Utah Juniper (*Juniperus osteosperma*):** Hardy tree with a twisted, rugged appearance, often found in upland deserts.

Moderate Water Needs & Varied Microclimates

1. **Aloe Vera (*Aloe barbadensis miller*):** Succulent with medicinal properties, can tolerate various microclimates if well-drained.
2. **Creosote Bush (*Larrea tridentata*):** Extremely drought-tolerant shrub with a distinctive smell and yellow flowers, adapts to different desert zones.

These plants illustrate a range of adaptations to desert life, including water conservation mechanisms like waxy coatings, deep root systems, and reduced leaf surfaces.

4.3 Creating a Water-Efficient Vegetable Garden

Vegetable gardens are often seen as water-intensive, but with thoughtful selection and innovative techniques, you can cultivate a lush, productive vegetable patch even in dry conditions. Starting with the right choice of vegetables is crucial. Opt for varieties known for their drought resistance, which often means they have adapted to thrive with less water. For example, consider planting artichokes, Swiss chard, or okra. These vegetables not only tolerate low water conditions but can also add unique flavors and textures to your culinary creations.

When you select vegetables, pay attention to their specific varieties as some have been bred specifically for drought tolerance. For instance, certain types of tomatoes and peppers have varieties that perform exceptionally well with minimal watering. It's also beneficial to explore local heirloom varieties that have naturally adapted to your region's climate over generations. These local adaptations make them particularly suited for thriving in specific conditions with less water. Engaging with local gardeners or visiting a nearby nursery can provide insights into the best varieties for your area.

Growing these vegetables effectively in dry conditions involves more than just careful selection; it also requires adopting specific cultivation techniques that conserve water. Spacing your plants appropriately plays a significant role in this. Too close, and they compete fiercely for every drop of water; too far apart, and the soil moisture evaporates quickly. Find the sweet spot where each plant has just enough room to grow without wasting valuable space that could retain moisture. Using mulch in your vegetable garden is another effective strategy. Organic mulches like straw or shredded leaves help retain soil moisture, keep the roots cool, and add organic matter to the soil as they decompose.

Timing your watering can also make a significant difference in how much water your vegetable garden requires. Watering early in the morning reduces evaporation and ensures that plants have adequate moisture throughout the day when they need it most. It can also help prevent fungal diseases that might otherwise thrive with evening watering. If manual watering isn't your style, consider setting up a drip irrigation system that targets the root zones directly, minimizing waste and ensuring that water goes exactly where it's needed.

Another technique is to use vertical gardening to maximize space and reduce water usage. Growing vegetables such as beans, cucumbers, and tomatoes upward rather than outward can not only save space but also reduce their exposure to soil pathogens and improve air circulation around the plants, which helps keep the leaves dry and reduces the risk of disease. Vertical structures can also provide shade for more sensitive plants or soil, further reducing water evaporation. Another advantage of vertical cultivation is it is conducive to interior spaces as well as outdoor locations.

When you implement these strategies, your vegetable garden becomes a testament to efficiency and productivity. It's not just about growing food; it's about nurturing a sustainable ecosystem right in your backyard. Each step, from selecting the right varieties to implementing water-saving cultivation practices, contributes to a garden that provides not only nourishment but also a sense of accomplishment and harmony with the environment. As you watch your vegetables thrive, each thriving plant stands as proof that with the right techniques, you can create an abundant garden that respects the limits of our natural resources.

Vegetable gardening in desert climates can be challenging due to the intense sun, low humidity, and infrequent rainfall. However, several vegetable species have been bred or are naturally suited to thrive in such conditions. Here's a list of at least 10 vegetable plants that are well-adapted to desert climates, including some specific breeds for arid conditions:

Plant Profiles: Key Desert Vegetable Plant Varieties

Growing vegetables in desert climates poses unique challenges, including extreme heat, intense sunlight, and minimal rainfall. However, some vegetables have been specially bred or are naturally suited to these conditions.

Here are some vegetable plants adapted for desert climates, including specific breeds and heirloom varieties known for their resilience in arid conditions:

1. **Artichokes:** Particularly the Imperial Star variety, which is adapted to handle hot, dry conditions and can be grown as an annual in desert climates.

2. **Tomatoes:** Heat-tolerant varieties such as Heatwave II, Solar Fire, and Desert Pride are specifically bred to endure the high temperatures typical of desert regions.

3. **Peppers:** Many pepper varieties thrive in heat, including Anaheim, Jalapeño, and Serrano. The Bhut Jolokia (Ghost Pepper) is also notably resilient in hot climates.

4. **Eggplant:** Varieties like Black Beauty and Florida Market are well-suited for hot, arid environments due to their heat tolerance.

5. **Swiss Chard:** This leafy green, especially the Bright Lights and Fordhook Giant varieties, is very heat-tolerant and can produce well throughout a desert summer.

6. **Cucumbers:** Armenian and Desert cucumbers are particularly adapted to withstand hot, arid conditions while still producing a good yield.

7. **Melons:** Watermelon and cantaloupe varieties, such as Desert King and Navajo Sweet, are bred for arid conditions, managing to produce fruit in dry environments.

8. **Okra:** A naturally heat-loving plant, okra varieties like Clemson Spineless and Red Burgundy thrive in hot, dry climates.

9. **Cowpeas (Black-eyed Peas):** These are extremely drought-resistant and perfect for arid regions; varieties like California Blackeye and Pink Eye Purple Hulled are particularly resilient.

10. **Squash:** Summer squashes like Zucchini, Yellow Crookneck, and Patty Pan perform well in heat, as long as they are watered adequately.

These plants can manage the rigors of desert gardening with appropriate care, such as using mulches to conserve soil moisture, providing some shade during the hottest parts of the day, and employing efficient watering methods like drip irrigation or ollas to maximize water use. This approach will help you maintain a productive vegetable garden even in challenging desert conditions.

4.4 Seasonal Care for Drought-Resistant Plants

Understanding how to cater to the seasonal needs of drought-resistant plants is crucial for maintaining their health and vigor throughout the year. Each season brings its own set of challenges and opportunities for gardening, and being aware of what your plants need as the seasons change can make a significant difference in their development and survival. As you nurture these resilient varieties, remember that their minimal water requirements do not exempt them from needing careful seasonal adjustments and observations.

Understanding Seasonal Needs

The cycle of seasons affects all living things, and plants are no exception. Drought-resistant plants, while hardy, still react to changes in temperature, light, and moisture. During spring, these plants exit dormancy and prepare for a period of active growth, which means they might need slightly more water and nutrients. Summer might bring about stress from higher temperatures and increased evaporation, requiring adjustments in watering and perhaps some shade. As autumn approaches, plants begin to prepare for dormancy; this is a good time to strengthen them against the coming cold. Finally, winter often means ensuring that they are protected from frost and excessive wetness, which could be detrimental, especially for those not native to cold climates.

Each of these seasons requires you to be observant and responsive to the subtle cues your garden provides about the well-being of your plants. This might mean adjusting water levels, providing frost protection, or simply letting the plants be as they prepare naturally for changes in the season.

Spring Preparation

Spring is a season of awakening and growth for many drought-resistant plants. It's a time when careful preparation can set the stage for a season of lush, vigorous growth. Start by gently pruning any dead or damaged branches, which helps stimulate new growth and maintains the plant's shape. Pruning also improves air circulation, which is crucial in preventing fungal diseases that thrive in damp, stagnant conditions.

Mulching is another critical spring activity. Applying a fresh layer of mulch helps conserve moisture, which is increasingly important as temperatures start to rise. It also suppresses weeds, which compete with your plants for water and nutrients. Choose a mulch that complements the needs of your drought-resistant plants—organic mulches like bark or cocoa hulls not only retain moisture but also break down over time, enriching the soil.

Summer Maintenance

Summer can be challenging for gardens, especially in arid climates. This season calls for diligent care and attention to ensure that your drought-resistant plants do not succumb to the stress of heat and reduced water availability. Monitoring your garden for pests and diseases becomes particularly important during these warm months. Pests are more active in summer, and early detection and management can prevent significant damage.

Adjusting watering practices may also be necessary. Although drought-resistant plants are adapted to survive with less water, extended periods of heat and dryness can still stress them. Increase water slightly, but always water deeply and less frequently to encourage strong root growth, which is better for long-term drought resistance. Also, consider setting up shade cloth over more sensitive plants to protect them from the harshest midday sun, which can sometimes be too intense even for drought-tolerant varieties.

Winter Protection

As the growing season winds down and temperatures drop, preparing your drought-resistant plants for winter is essential, especially in regions where frost or snow is common. Many drought-resistant plants are sensitive to frost, which can damage or even kill them. Employ frost protection strategies such as covering plants with frost cloth or moving container plants to more sheltered locations.

For those in containers, ensure that they have adequate drainage and are not sitting in waterlogged soil, which can freeze and harm the roots. Additionally, reduce watering in the winter months, as plants typically enter a dormancy phase and require much less moisture—over watering during this time can lead to root rot and other issues.

Navigating the care of drought-resistant plants through the seasons involves a blend of attentiveness and understanding—knowing when to intervene and when to let nature take its course. By tuning into the needs of your plants throughout the year, you ensure not only their survival but also their thriving, which contributes to the overall resilience and beauty of your garden.

As this chapter closes, you've gained insights into not only selecting the right drought-resistant plants but also nurturing them through the seasons. Each season's care routine builds upon the next, ensuring your garden remains robust year-round. This holistic approach to plant care, tailored to the unique challenges of each season, ensures your garden is not only a reflection of careful planning but also of dynamic, responsive gardening. Moving forward, the next chapter will delve into strategies for adapting gardening practices to dry climates, where understanding and working with your environment becomes the key to gardening success.

Chapter 5

Adapting Standard Gardening Practices to Dry Climates

Imagine transforming your lush, water-intensive garden into a sustainable oasis that thrives on minimal water. A gardener in a nonarid region can benefit from adopting dry gardening practices to conserve water, build more resilient plants, and reduce dependence on irrigation, important due to increasingly unpredictable weather patterns and droughts. Dry gardening also promotes healthier soil, as it encourages deep root growth and natural moisture retention, leading to a more sustainable and low-maintenance garden. This transformation requires a change in perspective and practice, a journey from traditional gardening to a philosophy where every drop of water is valued. In this chapter, we explore how you can make this transition smoothly, addressing the challenges and celebrating the successes along the way.

5.1 Transitioning From Traditional to Dry Gardening

Mindset Shift: Embracing Water Efficiency and Sustainability

The first step in transitioning to dry gardening is a shift in mindset. Traditional gardening often relies on abundant water use, with little thought to conservation or the environmental impact. Dry gardening, on the other hand, is rooted in the principles of sustainability and efficiency.

It challenges you to see the garden not just as a space of beauty but as an ecosystem where every element works in harmony, and every resource is optimized.

This shift begins with understanding the value of water as a precious commodity. In regions affected by drought or where water restrictions are in place, this understanding becomes crucial. But even in less arid areas, adopting water-efficient practices can have significant environmental benefits, including reducing runoff and preserving local aquifers. Think of water as a resource that needs to be carefully managed and conserved, and you'll start to see how your gardening practices can evolve.

Step-by-Step Transition: Gradually Integrating Dry Gardening Techniques

Transitioning to dry gardening doesn't have to be an overnight overhaul. A gradual integration of dry gardening techniques can be more sustainable in the long run, allowing you both to learn and adapt as you go. Start with small, manageable changes. For instance, you might begin by replacing water-thirsty plants with drought-resistant varieties in one section of your garden. As you become more comfortable with these plants and their needs, you can expand these changes to other parts of your garden.

Another effective strategy is to introduce mulching and efficient irrigation systems like drip irrigation or the use of ollas. These changes not only reduce water use but also improve soil health and plant resilience. Over time, these incremental changes will transform your garden into a model of dry gardening, without the overwhelm that might come with a complete immediate overhaul.

Overcoming Challenges: Practical Solutions for Common Issues

Transitioning to dry gardening is not without its challenges. One common issue is the adaptation of plants to new watering regimes. Plants that are used to frequent watering may experience stress when water is reduced. To mitigate this, reduce watering gradually, allowing plants to adjust slowly. Additionally, invest time in improving your soil's water retention capacity by adding organic matter or using soil amendments. This can make a significant difference in how well your garden adapts to less frequent watering.

Another challenge is the mindset of others who may be used to traditional gardening practices. Educating family members or community gardeners about the benefits of dry gardening can help overcome resistance. Share resources, offer tours of your transitioning garden, and discuss the long-term benefits, including reduced water bills and lower environmental impact.

Success Stories: Inspirational Examples to Motivate and Guide

Sometimes, seeing is believing. Throughout this transition, look to success stories for inspiration and practical insights. For instance, consider the story of a community garden in an urban setting that successfully transitioned to dry gardening. By using native plants, mulching, and efficient irrigation, they were able to reduce their water usage by 50% while increasing their garden's biodiversity, attracting more pollinators, and improving the aesthetics of the space.

Another powerful example is a homeowner in a suburban area who replaced their traditional lawn with a drought-tolerant landscape featuring native grasses and wildflowers. Not only did this change drastically cut down their water use, but it also provided a vibrant, wildlife-friendly yard that became a model for the neighborhood.

Textual Element: Reflection Section

Take a moment to reflect on your current gardening practices. Consider the amount of water you use and think about areas where you could reduce this usage. How might your garden benefit from a transition to dry gardening? What are your biggest concerns or challenges in making this change? Reflecting on these questions can help you set clear, achievable goals as you begin to integrate more sustainable practices into your gardening routine.

As you embark on this transformative process, remember that each small step contributes to a larger goal of sustainability and efficiency. With each plant you replace and each drop of water you save, you're not only creating a garden that's adapted to dry climates but also contributing to a more sustainable world. Whether you're dealing with the arid conditions of the desert or the occasional drought in temperate zones, the techniques and principles of dry gardening can empower you to create a thriving garden that respects and utilizes every resource thoughtfully and efficiently.

5.2 Soil Amendment Techniques for Better Water Retention

Improving the soil in your garden is akin to laying a strong, nurturing foundation for your home. Just as a robust foundation supports a house, well-amended soil supports the health and vigor of your plants, particularly in dry gardening where water efficiency is paramount. The texture, structure, and organic content of your soil play critical roles in how well water is retained and how easily it is available to your plants. Enhancing your soil not only improves water retention but also boosts the overall fertility and structure, creating an ideal environment for plant roots to thrive.

Amending your soil to enhance its water retention capabilities begins with understanding its current condition. If your garden's soil drains too quickly, it will benefit from amendments that increase its ability to hold water. Conversely, if your soil holds water for too long, creating a soggy environment, it's probably high in clay and needs amendments to improve drainage while not decreasing its ability to hold onto necessary moisture. The goal is to reach a balance where water is retained long enough to be used by the plant but not so long that it leads to root diseases or oxygen deprivation.

One of the most effective ways to improve soil structure for water retention is the incorporation of organic matter. Adding well-decomposed compost, aged manure, or leaf mold increases the organic content of the soil, which enhances its texture and fertility. These organic materials act like sponges, holding onto water and releasing it slowly to plant roots as needed. They also improve the soil's structure, increasing its porosity and aeration, which allows roots to penetrate more deeply and soil to drain more effectively while retaining essential moisture. This not only makes your garden more drought-tolerant but also reduces the frequency of watering needed, making your gardening efforts more sustainable and efficient.

In addition to traditional organic materials, consider using water-absorbing additives like biochar and vermiculite to further enhance water retention. Biochar, a form of charcoal used as a soil amendment, has a porous structure that can significantly improve soil's water-holding capacity. It not only retains water but also stores nutrients, releasing them slowly over time, which helps reduce the need for frequent fertilization. Vermiculite, a natural mineral that expands with heat, is another excellent amendment for dry gardens. Its lightweight, spongy texture helps loosen compacted soil, improving aeration and moisture retention. When mixed into the soil, vermiculite acts as a reservoir, absorbing water and nutrients and making them available to plant roots as needed.

Another transformative practice for enhancing soil water retention is adopting layering and no-dig gardening methods. No-dig gardening involves layering organic materials directly on top of the soil, which then decompose naturally, feeding the soil from above. This method not only minimizes soil disturbance, preserving the existing soil structure but also helps maintain higher moisture levels within the soil. Additionally, the no-dig method encourages the proliferation of earthworms and beneficial microbes, which are vital for healthy soil ecology. These organisms help create natural channels in the soil, improving its texture and ability to retain water.

Embracing these soil amendment techniques transforms your garden's ground into a reservoir of nourishment and hydration. As you continue to nurture and enrich your soil, you'll find that your garden becomes a more resilient, vibrant, and water-efficient space, capable of withstanding the vagaries of weather and the challenges of dry gardening.

5.3 The Role of Microclimates in Dry Gardening

Microclimates are small areas within your garden where the climate differs from the surrounding areas, and understanding them can be your secret weapon in creating a thriving dry garden. These varied pockets of warmth, moisture, and light aren't just anomalies; they're opportunities. By learning to identify and utilize these microclimates, you can strategically place plants where they are most likely to succeed, conserve water, and enhance the overall health of your garden.

Firstly, recognizing these microclimates involves keen observation over time. Start by noting how sunlight moves across your garden throughout the day. You might find areas that receive full sun all day and others that remain shaded. Temperature variations are also crucial; areas near walls might stay warmer, while low spots might collect cooler air at night. Even soil moisture can vary significantly within a small garden—slopes may lead to dry conditions at the top and moist conditions at the bottom. By mapping these observations, you begin to form a picture of your garden's unique microclimate landscape.

Once you've identified these microclimates, the next step is to leverage them to enhance your garden's efficiency and productivity. For instance, plants that thrive in full sun and need less water can be positioned in the warmer, drier parts of your garden, reducing the need for artificial watering. Conversely, areas that retain moisture and are cooler could be ideal for plants that require more water and less heat. This strategic placement not only ensures that your plants are more likely to thrive but also naturally reduces your garden's water demand.

Creating beneficial microclimates might sound complex, but it can be as simple as altering the landscape or introducing structures that modify environmental factors. For example, planting a row of trees or shrubs on the north side of your garden can protect against cold northern winds, creating a warmer microclimate. Similarly, installing a trellis or shade structure, or planting tall plants along the south side can provide necessary shade for heat-sensitive plants, reducing soil temperature and moisture evaporation. Even the color and material of your garden boundaries, like fences or walls, can influence the microclimate. Darker colors absorb more heat, which can be beneficial in cooler regions, while reflective materials might be better in very hot areas.

Managing microclimates effectively can lead to substantial water savings. For instance, by concentrating your watering efforts on cooler, shaded areas where evaporation rates are lower, you maximize the effectiveness of every ounce of water used. Drip irrigation systems can be tailored to the needs of different microclimates within your garden, delivering water precisely where and when it is needed. Mulching practices can also be varied—thicker mulch can be used in sunnier areas to prevent water loss, while less can be used in naturally moist areas.

Understanding and manipulating microclimates allows you to reduce the need for external inputs like water and fertilizers and increase the sustainability of your gardening practices. As you continue to observe and interact with these microclimates, you'll find that your garden not only survives but thrives, adapting naturally to the dry conditions that once seemed like a challenge.

5.4 Pest Management in the Dry Garden

In the unique ecosystem of a dry garden, managing pests effectively while conserving water and maintaining ecological balance presents a distinct set of challenges and opportunities. The dry conditions that help save water and support drought-resistant plants can also influence the types of pests that are likely to be encountered. Common pests in dry gardens often include spider mites and aphids, which thrive in hot, dry conditions and can quickly become problematic by stressing plants even further, making them more susceptible to disease and other pest infestations.

Integrated pest management (IPM) is a holistic approach that can be particularly effective in managing these challenges. IPM focuses on long-term prevention and control through a combination of techniques that are environmentally sensitive and sustainable. It starts with the proper identification of pests, understanding their life cycles and the role they play in the ecosystem. This knowledge allows you to implement targeted strategies that are more about managing pests than eradicating them completely, thus preserving the ecological balance within your garden. For example, introducing natural predators like ladybugs can keep aphid populations in check while installing insect netting during critical periods can prevent infestations without the need for chemical interventions.

Using natural and organic remedies is a cornerstone of IPM in dry gardens. These methods not only align with the sustainable ethos of dry gardening but also help to minimize water pollution and soil degradation that can result from synthetic chemicals. Neem oil, for instance, is a natural pesticide that works effectively against a variety of pests and is safe for use around pets and wildlife. Similarly, insecticidal soaps can target soft-bodied pests like spider mites without harming beneficial insects or adding harmful residues to the soil. These treatments require minimal application, reducing both water use and environmental impact.

Preventative practices play a crucial role in the IPM approach, helping to avoid major pest outbreaks before they start. Selecting plant varieties that are resistant to pests is a fundamental strategy. Many native plants, adapted to local conditions, tend to be more resistant to pests and diseases prevalent in that area. Proper garden hygiene, such as removing dead leaves and debris, eliminates potential breeding grounds for pests and reduces fungal diseases, which can be prevalent in areas that do experience occasional moisture. Additionally, proper watering techniques that avoid overhead sprinkling can help keep foliage dry and less hospitable to pests that thrive in moist conditions.

Implementing these IPM strategies does more than just control pests; it contributes to a healthier, more resilient garden ecosystem. By understanding and working with nature, you can maintain a productive dry garden that supports not only the plants but also the beneficial insects and wildlife that make a garden thrive. This approach enhances the beauty and productivity of your garden and also ensures it remains a sustainable and environmentally friendly space.

5.5 Fertilizing Strategies That Support Water Conservation

Understanding the intricacies of plant nutrition unveils a world where every nutrient plays a pivotal role in the growth and resilience of your garden. Plants require a balanced diet of nutrients to thrive, and this is particularly crucial in dry gardening where water is a scarce resource. Efficient use of fertilizers can significantly enhance plant health, making them more robust against drought and less dependent on frequent watering.

At the heart of effective fertilization is the choice of fertilizer. Organic fertilizers, such as compost, fish emulsion, and seaweed extracts, not only provide a slow-release source of nutrients but also improve soil structure and water retention. These natural fertilizers break down gradually, supplying a steady stream of nutrients to plants. This slow decomposition process encourages microbial activity, which enhances soil structure, allowing it to retain more moisture and nutrients. The enhanced soil structure ensures that water penetrates deeply into the soil, reaching the root zone where it's most needed and reducing surface evaporation.

In contrast to their synthetic counterparts, organic fertilizers are less likely to leach away, which means they remain in the soil longer, making them more available to plants and less likely to contaminate local water sources. This attribute is particularly important in dry gardening, where conservation and efficient use of resources are paramount. By choosing organic options, you're not only feeding your plants but also nurturing the soil ecosystem, supporting a cycle of growth and sustainability that benefits every aspect of your garden.

The timing and method of fertilizer application play critical roles in maximizing the effectiveness of these nutrients while conserving water. Ideally, fertilizers should be applied at times when plants can utilize them most efficiently, such as just before the peak growing season when plants are actively growing and can readily absorb nutrients. Applying fertilizers too early or late in the season can lead to wastage, as plants are not able to use the nutrients effectively, and they simply wash away.

Moreover, methods such as side-dressing or incorporating fertilizers into the soil near the plant roots can be significantly more effective than broadcasting them over a larger area. This targeted approach minimizes waste and ensures that nutrients are delivered precisely where they are needed. Additionally, watering lightly after applying fertilizer can help to integrate nutrients into the soil without washing them away, ensuring they reach the root zone where they are most beneficial.

In the broader context of environmental stewardship, reducing or eliminating the use of synthetic fertilizers can have profound benefits. Synthetic fertilizers often contain high levels of salts, which can build up in the soil over time, degrading soil quality and leading to increased water use. Furthermore, the production of synthetic fertilizers is energy-intensive and contributes to greenhouse gas emissions. By shifting towards organic fertilizers, you not only support the health of your garden but also contribute to a more sustainable and environmentally friendly gardening practice.

As we conclude this exploration of fertilizing strategies, remember that your choices have a broad impact, reaching beyond your garden. This chapter sets the stage for the next, where we will delve into technological advances in dry gardening, exploring innovative tools and techniques that can further enhance the sustainability and productivity of your garden. In this continuous journey of learning and adaptation, each step you take not only enriches your garden but also reinforces your commitment to nurturing a more sustainable world.

Make a Difference with Your Review

Unlock the Power of Giving

"Money can't buy happiness, but giving it away can."
–Freddie Mercury

Imagine helping someone you've never met without expecting credit. That person might be just like you once were—eager to make a difference but unsure where to start. Our mission is to make Dry Gardening accessible to everyone, but we can't do it alone.

Your review can help us reach more people. It takes less than 60 seconds but can make a lasting impact. By sharing your thoughts, you might help:

- One more small business thrive.
- One more entrepreneur support their family.
- One more individual find meaningful work.
- One more dream come true.

If you believe in helping others, take a moment to leave a review. It costs nothing but can change someone's life.

Simply scan the QR code or click the link below to leave your review:

https://www.amazon.com/review/review-your-purchases/?asin=B0DN7VV6W1

Thank you for being part of our community. Your support means the world to me, and I'm excited to help you achieve your gardening goals in the upcoming chapters.

Warm regards,
Richard Dreisdale

Chapter 6

Technological Advances in Dry Gardening

Imagine a garden where technology and nature intertwine seamlessly, turning age-old gardening struggles into a modern-day symphony of efficiency and harmony. In this chapter, we delve into the world of smart irrigation technologies that transform not just how we water our gardens but also how we interact with them. These innovations are not just about saving water—they redefine the gardening experience, making it more connected, controlled, and conscious, aligning perfectly with the ethos of dry gardening.

6.1 Smart Irrigation Controllers and Sensors

Precision Watering: The Art of Giving Your Plants Exactly What They Need

The core of smart irrigation lies in precision watering—providing your plants with the exact amount of water they need, exactly when they need it, no more, no less. Smart irrigation controllers and soil moisture sensors are the maestros in this delicate balance of water management. These devices use data from various sources, including weather forecasts and soil moisture levels, to adjust watering schedules automatically. This means your garden receives the right amount of hydration without any guesswork or waste.

For example, soil moisture sensors can detect the moisture level deep within the soil, ensuring that water is only applied when necessary and at optimal levels for plant health. This targeted watering approach prevents both under-watering and waterlogging which can stress plants and lead to root diseases. The precision of these systems allows you to maintain the perfect moisture balance in your soil.

Tech Integration: Smart Gardening at Your Fingertips

The integration of smart irrigation technologies with smartphones and home automation systems brings a new level of convenience and control to gardening. Imagine adjusting your garden's watering schedule from the comfort of your living room or receiving notifications on your phone when it's time to water or when adjustments are made based on the day's weather.

This connectivity not only makes garden management more straightforward but also more responsive. You can react in real time to changes in weather or soil conditions, adjusting settings as needed to ensure optimal growth conditions. The integration with home automation systems like Amazon Alexa or Google Home also allows for voice control, further simplifying the management of your garden's needs.

Water Savings: Quantifying the Environmental Benefits

The environmental benefits of using smart irrigation technologies in your garden are substantial. By optimizing water usage, these systems have reduced the amount of water used in many gardens by up to 50%.

Ease of Use: Democratizing Advanced Gardening Techniques

One of the most compelling aspects of smart irrigation systems is their user-friendly nature. These systems are designed to be intuitive, making them accessible to gardeners of all skill levels. Whether you're a tech-savvy gardener or someone who prefers a more hands-off approach, smart irrigation systems can be easily set up and maintained.

For instance, the Rachio 3 Smart Sprinkler Controller is renowned for its ease of installation and user-friendly interface. It offers a straightforward setup that integrates seamlessly with your existing irrigation system and connects to your home Wi-Fi network, allowing you to manage your watering schedule through a simple app. The app not only lets you control the watering times but also provides insights into your water usage and savings, making it easy to see the impact of your choices on both your garden and the environment.

Smart irrigation technologies represent a revolution in garden management, blending convenience with conservation. These tools empower you to maintain a lush, thriving garden that aligns with the principles of sustainability and efficiency, proving that even the ancient art of gardening can benefit from a touch of modern innovation.

6.2 Use of Aquaponics in Dry Gardening

System Basics: Integrating Fish Farming With Hydroponics

Aquaponics represents a remarkable synthesis of aquaculture and hydroponics, a symbiotic system where the waste produced by farmed fish supplies nutrients for plants grown hydroponically, which in turn purify the water. This closed-loop system is an ingenious solution for dry gardening, as it dramatically enhances water efficiency while providing you with both fresh produce and fish.

The heart of the aquaponics system lies in its ability to recycle water. In a typical setup, water from the fish tank, rich in organic waste, is pumped to the plant grow beds. Here, bacteria naturally present in the environment convert ammonia from fish waste first into nitrites and then into nitrates, which are excellent plant nutrients. The plants absorb these nutrients, effectively cleaning the water, which is then recirculated back to the fish tank. This cycle is continuous, ensuring that water usage is minimized.

Water Efficiency: Revolutionizing Resource Use in Gardening

Aquaponics shines in its ability to use up to 90% less water than traditional soil-based gardening. In dry gardening contexts, this level of water efficiency is transformative, enabling lush, productive gardens in areas where water restrictions or scarcity might otherwise limit gardening activities. The closed-loop nature of aquaponics means that water loss occurs only through transpiration and evaporation, significantly less than the water used in traditional gardening, where much can be lost to soil absorption or runoff.

The efficiency extends beyond water to space and nutrient use. Aquaponics systems can be set up vertically, maximizing space usage—an ideal solution for small gardens or urban environments where space is at a premium. Moreover, the nutrient cycle in aquaponics is nearly closed: fish feed is the primary input, and the system circulates and recycles the resulting nutrients through the plants, requiring fewer external nutrients and almost no soil, which can be both a resource constraint and an environmental load in terms of production and degradation.

Sustainability Aspects: Nurturing an Ecosystem in Your Backyard

The sustainability of aquaponics is multifaceted, encompassing environmental, economic, and social dimensions. Environmentally, aside from dramatically reducing water use, aquaponics eliminates the need for chemical fertilizers, which can be harmful to both the environment and human health.

Economically, aquaponics can reduce grocery bills by providing a steady supply of fresh produce and fish, high-cost items in many areas. Furthermore, the low operational costs associated with water and fertilizer use, coupled with the high efficiency of the system, can make it a financially viable option even for small-scale gardeners. Socially, aquaponics provides an opportunity for community building and education. It's a compelling model of sustainable food production that can be used in schools, community centers, and homes, providing a platform for education about ecology, nutrition, and resource management.

Setup and Maintenance: Creating and Sustaining Your Aquaponics Garden

Setting up an aquaponics system doesn't have to be overly complex, but it does require some basic components: a fish tank, grow beds, a water pump, and plumbing to connect these elements. The cost of setting up can vary depending on the scale and the materials used. For a basic home system, investment might be in the range of a few hundred dollars, but it's important to consider this against long-term savings on groceries and water bills.

Maintenance involves monitoring the health of both the fish and the plants, ensuring that the water pH and nutrient levels are balanced and that the system is free from clogs and leaks. While this might sound daunting, many aquaponics gardeners find the daily interaction with their system deeply rewarding—not just for the tangible yields of fresh produce and fish, but for the connection to a miniature ecosystem. Observing the interactions between plants and fish can provide insights into ecological and biological processes, making aquaponics not just a gardening method but a continual source of learning and discovery.

As you delve into the world of aquaponics, you embark on a path that is about more than just gardening; it's about fostering a sustainable relationship with your environment, where waste becomes a resource and gardening becomes a loop of continuous growth and regeneration. Whether you are a seasoned gardener or new to the green-thumb world, aquaponics offers a dynamic way to engage with your food and your ecosystem, making every drop of water work multiple times for you and your garden.

6.3 Vertical Gardening: A Solution for Small Spaces

When you live in an urban setting or have limited outdoor space, it might seem like having a lush garden is out of reach. However, vertical gardening provides a creative and effective solution to cultivate a vibrant patch of greenery without requiring the traditional expanses of garden space. By taking advantage of vertical space, from balcony railings to sunny walls, you can transform even the narrowest of areas into productive gardens. This method is particularly well-suited for dry gardening, as it allows for optimal use of limited resources, including water.

Vertical gardening is all about maximizing the potential of vertical surfaces. Techniques such as using trellises, wall-mounted planters, or even repurposed items like pallets and hanging pockets, can all serve as foundations for growing plants upwards instead of outwards. These structures not only save on ground space but also can create microclimates that may benefit certain plants. For instance, a wall that receives ample morning sunlight but is shaded during the hottest part of the day is an ideal location for plants that need light but are prone to wilting under intense heat.

The choice of structure depends largely on what you intend to grow and the conditions of your space. Freestanding trellises or tower gardens are perfect for vine plants like tomatoes and beans, while wall-mounted systems can be great for herbs and small vegetables. These installations can be as simple or complex as your budget and skills allow. A basic trellis can be crafted from inexpensive materials like bamboo or upcycled metal rods, offering a functional and eco-friendly option, while more sophisticated living walls might involve an initial higher investment but pay off in terms of visual impact and space efficiency.

Selecting the right plants for a vertical garden is crucial, especially when considering water requirements. In dry gardening, the goal is to choose varieties that thrive with minimal water. Succulents, with their low water needs and shallow root systems, are natural fits for vertical gardens. Similarly, Mediterranean herbs like rosemary and thyme not only adapt well to vertical setups but also require less frequent watering. When planning your plant selection, consider the microclimate created by your vertical structure and how it affects moisture retention and sunlight exposure. Plants at the top might dry out faster and could be more suited to drought-tolerant species, while those lower down might retain moisture longer and could be suitable for slightly thirstier plants.

Efficient watering strategies are key in vertical gardening, especially under the constraints of dry gardening. Drip irrigation systems are ideal for these setups as they deliver water directly to the base of each plant, minimizing waste and evaporation. These systems can be easily integrated into vertical structures and, when automated, provide the added benefit of delivering just the right amount of water at the right time, further enhancing water efficiency. For those on a tighter budget or preferring a more hands-on approach, self-watering containers offer an alternative. These containers have reservoirs at the bottom that allow plants to absorb water as needed, reducing the frequency of watering and ensuring that water use is optimized for each plant's needs.

Setting up a basic vertical garden is surprisingly affordable and can be a fun weekend project. Starting with a simple frame or wall panel, which can range from $50 to $200 depending on size and materials, you can create the backbone of your garden. Add in some containers or pots, tailored to fit your structure, and some high-quality potting soil or a suitable growing medium, and you're almost ready to plant. Watering systems can vary in complexity and cost, but a basic drip setup or a few self-watering pots will ensure your plants stay hydrated efficiently. Finally, selecting seeds or starter plants that match your setup and climate will set your vertical garden on the path to success.

In embracing vertical gardening, you are not just making the most of limited space; you are also committing to a sustainable form of gardening that prioritizes water efficiency and adaptability. The setup costs, while variable, are an investment in creating a garden that brings beauty, biodiversity, and a bit of food security to urban and small-space dwellers alike. As you watch your vertical garden grow, flourishing in a harmony of form and function, you'll find that this innovative approach to gardening is not just about growing plants up a wall—it's about elevating the potential of every square inch of your living space.

6.4 Hydrozoning: Grouping Plants by Water Needs

In the realm of dry gardening, where every drop of water counts, the concept of hydrozoning emerges as a strategic masterpiece in garden planning. Hydrozoning is the practice of grouping plants according to their water needs, ensuring that each zone within your garden is irrigated according to its specific requirements. This method not only maximizes irrigation efficiency but also significantly reduces water waste, tailoring your gardening efforts to be as environmentally friendly as they are effective.

Imagine your garden as a series of concentric circles, each with its own unique thirst level. In the innermost circle, you might place high-water-use plants that are lush and verdant, requiring more frequent watering. As you move outward, the circles encompass plants that demand progressively less water, ending with drought-tolerant plants that thrive on minimal hydration. This layout allows you to customize your watering schedule based on the needs of each zone, avoiding the inefficiency of watering all plants at the same rate regardless of their requirements.

The benefits of hydrozoning extend beyond just water savings. By grouping plants with similar water needs, you create a more homogenous environment, which can lead to better plant health and reduced maintenance. Plants that require similar amounts of water are more likely to thrive when placed together, as they compete less for moisture and nutrients. This not only makes your garden more resilient but also reduces the time and effort you need to spend on maintenance. Furthermore, hydrozoning can help prevent overwatering, a common issue that can lead to root rot, fungal growth, and other water-related plant diseases.

To implement a hydrozoned garden layout, start by assessing the water needs of your plants. This might involve some research into their specific moisture requirements or, if you're working with native plants, observing their performance in their natural habitat. Once you have a clear understanding of these needs, sketch a plan of your garden, dividing it into zones based on these criteria. Each zone should correspond to a different irrigation regime, which can be as simple as manual watering at different frequencies or as sophisticated as an automated irrigation system with zone-specific settings.

Optimizing irrigation for each hydrozone involves more than just adjusting the frequency of watering. It's also about tailoring the irrigation method to suit the needs of the zone. For example, drip irrigation is perfect for zones with moderate to low water requirements, as it delivers water directly to the roots, minimizing waste. For higher water-use zones, you might consider soaker hoses that provide more extensive coverage. The key is to match the irrigation method to the water needs of the plants, ensuring that each plant receives just the right amount of water.

The strategic beauty of hydrozoning lies in its simplicity and its profound impact on garden sustainability. As you water each zone according to its specific needs, you'll find that your garden not only survives but thrives, becoming a testament to the effectiveness of thoughtful, environmentally conscious gardening.

In this chapter, we have explored the cutting-edge technological advances that enhance dry gardening, from soil moisture sensors to Hydrozoning that optimizes watering efficiency. Building on this foundation, let's shift our focus to practical applications that you can implement in your own garden. The following chapter delves into several do-it-yourself projects that will help you maximize the potential of your dry garden. These projects are designed to enhance the technological tools we've discussed, ensuring your garden remains sustainable and thrives.

Chapter 7

DIY Projects for the Dry Garden

Imagine transforming your garden into an oasis of efficiency and sustainability, where every element works harmoniously to conserve water and enhance plant health. In this chapter, we dive into the world of DIY projects that not only add character and functionality to your garden but also align perfectly with the principles of dry gardening.

7.1 Drip Irrigation Systems for the Home Garden

DIY Drip Irrigation System

Installing a DIY drip irrigation system using a garden hose with punched holes is the most cost-effective way to efficiently water your garden. This simple method can help you maintain consistent moisture to your plants, reducing water evaporation and runoff and the effort required for hand watering. Here's a step-by-step guide on how to set up this system:

Materials Needed

- Garden hose (length depending on your garden size)
- Nail or ice pick (for punching holes)
- Hose cap or plug (to seal the end of the hose)
- Hose connector (to attach the hose to your water source)

- Optional: Hose timer (for automatic watering)

Steps to Install

1. **Plan Your Layout:**

    - Before modifying the hose, lay it out in your garden along the rows of plants where you want the water to drip. This will help you determine the best places to punch holes and ensure that the hose covers all necessary areas.

2. **Punch Holes:**

    - Use a nail or ice pick to carefully punch holes in the hose. The holes should be small and spaced according to the plants' needs. Typically, placing holes every six inches works well.
    - Ensure the holes are on the side of the hose that will be facing the roots of the plants to direct the water flow accurately.

3. **Seal the End of the Hose:**

    - At the end of your hose line, attach a hose cap or plug to prevent water from pouring out the end. This ensures that all water is directed through the punched holes.
    - You can also double the hose over and secure with a cable tie.

4. **Connect to Water Source:**

    - Attach the hose connector to the other end of the hose.
    - Connect this end to your outdoor tap or existing irrigation system.
    - Ensure all connections are secure to prevent leaks.

5. **Test and Adjust:**

    - Turn on the water to a low flow. Watch to see how the water drips out of the holes and adjust the water pressure as necessary. The goal is to provide a slow, steady drip that gives the plants adequate moisture without runoff.
    - You might need to adjust the size or spacing of the holes based on how the water is distributed.

6. **Secure the Hose:**

    - Use garden staples or hooks to secure the hose to the ground if needed. This prevents the hose from shifting and ensures that the water is delivered right where your plants need it.

7. **Optional—Install a Timer:**

    - For convenience, consider installing a hose timer at the faucet. This device can automate watering, ensuring your garden receives water at consistent times, which is better for plant health and water conservation.

Maintenance

- Regularly check the system for clogged holes or leaks.
- During freezing temperatures, drain and store the hose to prevent damage.

This DIY drip irrigation system is an excellent way to maintain your garden's health while conserving water. It can be adjusted and expanded as your garden grows and can also serve as a good foundation for more sophisticated irrigation setups in the future.

A Small System With Commercial Components

If you have a modest budget, a simple drip system using commercially available 1/2-inch tubing for main lines and 1/4-inch tubing for runs to individual plants or small groups of plants with inexpensive elbow and splice fittings available through the internet or at local garden supply stores offers numerous advantages over the most cost-conscious method above in a small residential garden. Such a system allows for the precise and predictable delivery of water through calibrated emitters to plant roots. This system can also be automated with timers ensuring that plants receive water at the most beneficial times, typically during cooler parts of the day. Inexpensive filters and pressure regulators are recommended for these systems.

Overall, while more cost-intensive than a DIY hose with holes, the use of a structured tubing system with adjustable emitters leads to more precise regulation. This makes it a wise investment for gardeners seeking a higher level of reliability and consistency in their watering practices. Additionally, design and installation advice is almost always available at the local or internet parts supply source.

This system also requires regular maintenance.

- Regularly check the system for leaks or clogged emitters.
- Drain all the lines during freezing temperature.

7.2 Building Your Own Olla Irrigation System

Cost-Effective Irrigation

Ollas (oh-yahs) are unglazed terracotta pots buried in the garden to facilitate slow, deep watering directly at the roots, where plants need it most. This method of irrigation is not only ancient but has stood the test of time as one of the most efficient ways to water plants, particularly in dry, arid climates. By using ollas, you minimize evaporation and runoff, ensuring that almost every drop of water is used effectively by your plants.

The beauty of using ollas lies in their simplicity and cost-effectiveness. Unlike high-tech irrigation systems, ollas can be easily made or purchased at a low cost and installed in your garden with minimal tools or prior experience. The process involves filling the olla with water, which then seeps slowly through the porous clay walls directly into the soil. This method provides a steady supply of moisture to plant roots, encouraging deeper root growth and reducing the frequency of watering needed.

Step-by-Step Guide

Creating your own olla irrigation system is a rewarding project that can be completed in just a few steps:

1. **Select Your Ollas:**
    - Choose terracotta pots that are unglazed, as these are porous and perfect for seeping water into the soil. You can find them at most garden centers or craft stores.
    - Ensure they have a wide opening for easy filling and a snug-fitting lid to prevent evaporation and deter pests. Be sure to plug any drainage holes in the pot to ensure that water releases slowly through the porous clay walls and not through large openings.
2. **Prepare the Garden:**
    - Identify where your plants need water the most and where the ollas will be most effective.
    - Dig a hole deep enough so that when the olla is placed inside, the neck of the pot remains just above the soil surface.
3. **Install the Ollas:**
    - Place the olla in the hole and backfill with soil, ensuring it's snug and stable. The top should remain exposed for easy refilling.

- Repeat this process throughout your garden, spacing ollas according to the water needs of different plants. An eight-inch pot will provide water effectively within a two-foot radius around it.

4. **Fill and Maintain:**

- Ollas are remarkably low-maintenance but do require occasional checks to refill water and ensure that no debris blocks the opening. Over time, the clay may develop a patina or algae on the surface, which is perfectly normal and does not affect its function. However, it's a good practice to clean your ollas periodically, removing any algae or mineral deposits to maintain optimal porosity in the clay. This can be done using a stiff brush and clean water, scrubbing gently to avoid damaging the pot.

- Maintaining ollas for long-term use is equally straightforward. During the off-season, it's wise to dig up and empty your ollas, giving them a thorough cleaning before storing them in a dry place. This prevents damage from freezing temperatures and extends the life of each olla. With proper care, ollas can be used for many gardening seasons, making them a cost-effective and environmentally friendly option for moisture management in your garden.

Plant Health

The slow and steady watering method provided by ollas promotes healthier plant growth in several ways. It encourages roots to grow deeper into the soil, seeking moisture from the olla, which in turn helps stabilize plants and protect them from drought conditions. This deep root system is crucial for long-term plant health, as it accesses nutrients from deeper soil layers and ensures stability against winds and erosion. Moreover, because the water goes directly to the roots, it reduces the moisture on the leaves, which can lead to fungal diseases and pest infestations.

By embracing the olla irrigation system, you're not just giving your plants an efficient water source—you're engaging in a time-honored gardening tradition that respects the earth's resources while nurturing your garden. This project is a testament to the fact that sometimes, the simplest solutions are the most effective, blending history, science, and art into your garden practices. As you watch your garden respond with vigor and vitality, you'll find that these ancient vessels bring not just moisture but a new depth of connection between you and your garden.

7.3 Constructing a Rain Garden for Water Runoff Management

Rain gardens are not just beautiful additions to your landscape; they are functional tools designed to tackle the issues of runoff and water waste—challenges that are particularly pronounced in areas prone to heavy rains or in urban environments where impermeable surfaces dominate. By strategically capturing runoff, rain gardens reduce the risk of flooding and soil erosion, channeling water away from structures and into the ground where it can do the most good. This method not only alleviates the strain on drainage systems but also enhances the availability of water for your garden during dry spells.

Creating an effective rain garden starts with understanding its basic structure and functionality. Unlike a typical garden bed, a rain garden is a shallow depression, which is designed to collect and filter runoff from roofs, driveways, and other hard surfaces. The idea is to slow the water down and funnel it into the ground, rather than letting it flow unimpeded into storm drains. By doing this, you not only prevent the pollutants in runoff from reaching larger bodies of water but also allow it to percolate through the soil, naturally filtering it and recharging the groundwater.

Plant Recommendations

The choice of plants is critical in a rain garden. You need species that can tolerate both wet and dry conditions—wet when the garden fills during a rainstorm, and dry when it empties a few days later. Native plants are typically the best choices because they're adapted to local climate conditions and require less maintenance. Look for plants that are known for their deep root systems, which help them withstand periods of both saturation and drought, while also enhancing the soil's structure and drainage capabilities.

For the water-tolerant zones at the center of the rain garden, consider plants like Joe-Pye weed, cardinal flower, and blue flag iris. These plants thrive in standing water and help filter pollutants. As you move to the edges of the garden, which may dry out more quickly, switch to drought-tolerant natives such as coneflower, black-eyed Susan, and switchgrass. These plants help stabilize the edges of the garden and provide habitat and food for local wildlife, adding to the ecological benefits of your rain garden.

Building the Rain Garden

1. **Select the Location:**
    - Choose a site at least 10 feet away from your home to prevent any risk of water damage.
    - The area should be where water naturally accumulates or where downspouts can be directed.
    - Avoid areas over septic systems or utility lines.
2. **Determine the Size:**
    - The size of the rain garden depends on the drainage area (roof, driveways, etc.) that will feed it.
    - A general rule is for the rain garden to be about 20–30% of the size of the area draining into it.
3. **Test the Soil:**
    - Conduct a soil infiltration test by digging a small hole, filling it with water, and observing how quickly it drains.
    - Ideally, the water should drain within 24 hours. Sandy soils drain faster than clay soils.

4. **Design the Layout:**

    - Shape the garden with a slight depression to hold water. A typical depth is 4–8 inches.

    - Plan for overflow by setting a path for excess water to exit, ideally leading to another permeable area.

    - Incorporating a berm—a small, raised barrier—on the downhill side of the garden can help retain water, ensuring it has enough time to soak into the soil.

5. **Excavate and Amend Soil:**

    - Remove the top layer of grass and soil to create the depression.

    - Mix the excavated soil with compost to improve infiltration and support plant growth.

6. **Select and Plant Appropriate Vegetation:**

    - Choose native plants that are adapted to your local climate and soil conditions.

    - Include a mix of deep-rooted plants to absorb water and nutrients. Consider plants that can tolerate both wet and dry conditions.

7. **Mulch and Water:**

    - Apply a layer of mulch to conserve moisture and suppress weeds.

    - Water the plants regularly until they are established.

8. **Maintain the Garden:**

    - Regular maintenance includes weeding, replanting as necessary, and checking the drainage features to ensure they remain functional.

Environmental Impact

The environmental impact of rain gardens extends beyond flood control and water filtration. By creating a diverse habitat, you're supporting local biodiversity, offering a refuge for birds, butterflies, and beneficial insects. This biodiversity is crucial for pollination and for maintaining healthy plant communities. Furthermore, rain gardens contribute to the reduction of heat in urban areas, countering the heat island effect that can make cities several degrees warmer than their surrounding areas.

As you plan your rain garden, consider its role not just in your landscape but as a part of the broader ecosystem. Each rain garden contributes to the health of the environment, providing benefits that extend beyond the confines of your yard. Whether you're a seasoned gardener or a novice looking to make a positive impact, constructing a rain garden offers a unique opportunity to blend functionality with environmental stewardship, creating a space that is as beneficial as it is beautiful. As you watch your rain garden mature and evolve, it becomes a dynamic testament to the role that thoughtful landscaping can play in sustainable living.

7.4 Building Shade Structures

Creating a thriving garden in arid desert regions is challenging due to the extreme heat, intense sunlight, and limited water. Shade structures provide a cooler microclimate, reduce water evaporation, and protect plants from sunburn and heat stress. This guide covers the design, construction, installation, and maintenance of shade structures for dry gardens in desert areas.

Design

1. Purpose and Functionality

    - Shade Coverage: Assess the level of shade needed for your plants, ranging from partial to full shade.

- Ventilation: Ensure proper airflow to prevent mold and mildew and keep plants healthy.
- Aesthetic Appeal: Integrate design elements that complement the garden's look and environment.

2. Materials

- Frame: Choose durable materials like galvanized steel, aluminum, or treated wood for harsh desert conditions.
- Shade Fabric: Use UV-resistant, breathable fabrics like polyethylene or polypropylene, available in various densities (30%, 50%, 70%).
- Anchoring: Secure the structure with strong anchoring systems, such as concrete footings or deep-set posts, to withstand strong winds.

Construction

1. Site Preparation

- Location: Select a site with adequate sunlight, considering sun direction and intensity. An east-west orientation is often best.
- Soil Preparation: Clear debris and level the ground for a stable foundation.

2. Frame Assembly

- Foundation: Dig holes and pour concrete footings for the posts. Allow the concrete to cure fully.
- Structure Assembly: Assemble the frame per design specifications, ensuring all joints are secure.

3. Installing Shade Fabric

    - Measuring and Cutting: Measure and cut the shade cloth to fit the frame, leaving extra fabric for adjustments.
    - Attaching Fabric: Secure the fabric to the frame using clips or zip ties, keeping it taut but not overly stretched.

Installation

1. Positioning

    - Optimal Placement: Position the structure to provide the desired shade throughout the day.
    - Anchoring: Ensure all anchoring points are secure.

2. Adjustments

    - Height and Angle: Adjust the height and angle as needed for optimal shading and airflow.
    - Accessibility: Ensure the structure does not block pathways or garden access.

Maintenance

1. Regular Inspections

    - Structural Integrity: Periodically check the frame for wear, rust, or damage.
    - Tighten any loose connections and replace damaged parts.
    - Shade Fabric Condition: Inspect the shade cloth for tears, fraying, or UV degradation, and promptly repair or replace damaged sections.

2. Cleaning

- Shade Fabric: Clean periodically with mild soap and water to remove dust and debris. Avoid harsh chemicals.
- Frame: Wipe down the frame and check for corrosion or damage, addressing issues immediately.

3. Seasonal Adjustments

- Winter Storage: Remove the shade cloth during cooler months to prevent storm damage.
- Reinstallation: Reinstall the shade fabric in spring, ensuring it is securely attached.

Implementing shade structures in a dry garden significantly enhances plant health and growth in desert environments. By carefully designing, constructing, installing, and maintaining these structures, gardeners can create a more hospitable microclimate that mitigates the challenges posed by extreme heat and sunlight.

7.5 Creating a Drought-Resistant Container Garden

Container gardening represents a fantastic avenue for cultivating a lush and vibrant garden inside or outdoors, especially when space is at a premium or when you seek greater control over the growing environment. This gardening style offers significant flexibility, allowing you to create a portable oasis that can be adjusted according to seasonal conditions and personal preferences. Particularly in urban settings or smaller residences like apartments with balconies, container gardening emerges as a practical solution, providing an accessible route to green spaces that are both beautiful and functional.

One of the most compelling advantages of using containers is the ability to precisely manage the soil and water conditions of your plants. Unlike traditional garden beds, containers prevent issues related to soil compaction and contamination, and they significantly reduce the risk of soil-borne diseases. This controlled environment allows you to tailor the soil mixture to suit the specific needs of different plants, optimizing conditions for growth and resilience. For instance, cacti and succulents thrive in fast-draining soil mixes which might be difficult to maintain in open garden beds. Containers make it easier to create these specialized conditions, ensuring your plants not only survive but thrive.

Choosing the right type of container is crucial for the success of your drought-resistant garden. Materials can vary widely—from traditional clay and terracotta pots to modern plastics and recycled materials. Each has its benefits and challenges. Terracotta and clay, for example, are porous, allowing for better air circulation to the roots and more natural water evaporation from the soil. However, they can require more frequent watering as a result. Plastic pots, on the other hand, retain moisture longer and are lighter and sometimes more cost-effective, but they might not offer the same breathability. In choosing the right container, consider the water needs of your plants and the climate conditions of your area. Ensure that whatever type you select, there is adequate drainage. Drainage holes are essential to prevent waterlogging, which can lead to root rot and fungal diseases.

When it comes to soil and plant choices, selecting the right mix and species is key to cultivating a thriving container garden in dry conditions. A well-draining soil mix, typically one part perlite or coarse sand to two parts potting soil, will ensure that water does not sit around the roots, which is particularly crucial in drought-resistant gardening. This type of mix not only prevents overwatering but also allows the roots to breathe, reducing the risk of disease. When selecting plants, opt for those that naturally thrive in dry conditions. Species such as lavender, sedum, and portulaca not only tolerate low-water conditions but also add vibrant colors and textures to your garden. Herbs like rosemary and thyme are also excellent choices, as they are naturally adapted to Mediterranean climates where dry conditions prevail.

Building a DIY Container Garden

Creating a small and inexpensive container garden using objects around your home is a great way to repurpose materials and add a bit of greenery to your space. Here's how you can do it:

1. **Select Containers:**
    - Old Kitchen Items: Things like colanders, tea kettles, and old pots make excellent plant containers. The holes in colanders are perfect for drainage.
    - Plastic Bottles and Containers: Cut open large soda bottles, milk jugs, or detergent containers. Be sure to make a few holes in the bottom for drainage.
    - Tin Cans: Empty soup or bean cans can be washed and used. Drill or punch holes in the bottom for water drainage.
    - Wooden Boxes or Crates: If you have any old wooden boxes or wine crates, these can be great for larger container gardens. Line them with plastic with holes for drainage to protect the wood.
2. **Prepare the Containers:**
    - Clean all containers thoroughly to remove any harmful residues.
    - If needed, paint or decorate the containers to fit your aesthetic taste. Use non-toxic paint to ensure it's safe for plants.
3. **Choose Your Soil and Plants:**
    - Use potting soil from a garden store, or make your own mix with equal parts peat, perlite, and compost.
    - Choose plants that are suitable for the container size and the available sunlight. Herbs, succulents, and small flowers are great for smaller containers.
4. **Planting:**
    - Fill the container with soil, leaving about an inch at the top for watering.
    - Dig small holes and place your plants inside. Cover the roots with soil and press down lightly.

- Water the plants gently to settle the soil around the roots.

5. **Placement and Care:**
    - Place your containers in a spot where they will receive the appropriate amount of sunlight based on the plants' needs.
    - Water the plants according to their specific requirements. Container plants often need more frequent watering because they dry out faster.
    - Fertilize occasionally with an appropriate fertilizer to ensure healthy growth.
6. **Ongoing Maintenance:**
    - Monitor your plants for pests and diseases. Treat with natural remedies like neem oil or insecticidal soap if needed.
    - Prune dead or overgrown parts to promote healthy growth and appearance.

With these steps, you can transform everyday objects into a charming, eco-friendly container garden that brightens up your home and offers the joy of gardening, even in small spaces.

Maintaining a container garden in dry conditions does require some attention, particularly to watering and mulching practices. While it's crucial to conserve water, it's equally important to ensure that plants receive enough moisture to thrive. A deep watering method, where water is applied slowly and deeply to encourage roots to grow downward rather than staying near the surface, is most effective. This method helps plants develop a resistance to drought and reduces the frequency of watering needed. Additionally, mulching your containers with organic materials such as bark or coco coir can help retain moisture, regulate soil temperature, and add a finished look to your plantings. This practice not only conserves water but also adds a layer of organic matter as the mulch breaks down, further enriching the soil.

As you delve into the world of container gardening, you'll discover that it offers a unique blend of convenience, control, and creativity. It allows you to bring nature into the most confined spaces, transforming balconies, patios, and windowsills into verdant landscapes that defy the dry conditions often seen in urban environments. Whether you are an experienced gardener or a novice green thumb, container gardening provides a versatile platform for cultivating a diverse array of plants, fostering a closer connection to nature even in the heart of the city. As you tend to your containers, adjusting soil mixes, watering deeply, and watching your plants flourish, you'll find that this form of gardening doesn't just beautify your space—it enhances your daily life, providing a lush, green sanctuary that is both a retreat and a testament to the possibilities of innovative gardening.

7.6 Rainwater Harvesting

The basics of setting up a rainwater harvesting system are straightforward. Begin with the collection area, typically your house roof. Gutters channel the water into downspouts where it can be directed into storage systems such as barrels or cisterns. These storage containers can be as simple as repurposed barrels placed under downspouts or as complex as large cisterns integrated into your garden's irrigation system. The critical aspect is ensuring they are covered to prevent debris entry and mosquito breeding. Ideally, the system should also include a first-flush diverter—a device that directs the first flow of rainwater away from the storage container. This water, which may carry particles from the roof or gutters, is diverted, ensuring that the water stored is cleaner.

Integrating harvested rainwater into your garden's irrigation can be a seamless process. For smaller garden setups, a gravity-fed system from a raised barrel can suffice. You can use a spigot at the bottom of the barrel to fill watering cans or connect a hose for direct watering. For more extensive systems, consider connecting your rainwater storage directly to a drip irrigation system. This setup might require a pump to ensure adequate water pressure, but the overall savings and efficiency in water usage will be well worth the initial setup effort. This integration not only reduces your reliance on municipal water systems but also provides your plants with soft, chemical-free water, which can be better for plant growth and soil health.

Navigating the legal and safety considerations of rainwater harvesting is crucial. Most regions support the use of collected rainwater for gardening, recognizing its environmental benefits, but local regulations may vary, particularly in areas where water rights are strictly managed. It's essential to check with your local water authority or environmental agency to ensure compliance with any regulations regarding collection and use. Safety considerations primarily involve ensuring that your system does not contaminate the stored water or become a breeding ground for pests. Regular maintenance of gutters, downspouts, and storage containers will keep the system safe and efficient. See Chapter 3 for additional information on rainwater harvesting.

7.7 Greywater System for Garden Use

Sustainable Water Reuse

Imagine repurposing the water from your shower, sinks, or washing machine to nourish your garden. This is the essence of greywater recycling—a smart, sustainable strategy that significantly enhances the efficiency of water use in your home. Greywater, which is all wastewater generated in your home except from the toilets, presents an opportunity to reduce the consumption of freshwater by diverting it for garden irrigation. This approach not only saves you money on water bills but also reduces the strain on municipal water treatment systems, aligning your gardening practices with broader environmental sustainability goals.

The concept of greywater recycling might seem complex, but at its core, it's about reusing water that's already been paid for and used once. Every time you shower or wash dishes, you're using water that's perfectly suitable for irrigation with some basic filtering. This water typically contains traces of dirt, food, grease, and household cleaning products, but with proper management, it can be safely used to irrigate non-edible plants. This practice not only provides a regular supply of water to your garden but also helps in maintaining moisture levels in the soil, crucial for gardens in dry climates where water scarcity can limit gardening activities.

Integrating a greywater system into your gardening routine requires an understanding of both its benefits and the responsibilities it entails. One of the biggest advantages is the reduction in freshwater usage, which conserves an important resource and can lead to significant savings. Additionally, using greywater for irrigation helps in recharging the groundwater in your area, contributing to a healthier local ecosystem. It's a proactive step towards a more sustainable lifestyle, where the reuse of resources becomes a daily practice, reflecting a commitment to environmental stewardship.

Legal and Safety Considerations

Before installing a greywater system, it's crucial to familiarize yourself with the local regulations that govern its use. Many regions have specific codes and guidelines designed to ensure that greywater systems do not pose a risk to health or the environment. These regulations can vary widely between different areas, so contacting your local water authority or environmental agency for guidance is a good first step. They can provide information on permissible uses of greywater, required treatment processes, and any necessary permits.

Safety is paramount when dealing with greywater because it can contain bacteria and chemicals from household products. To minimize risks, it's recommended to use greywater only for irrigating ornamental plants or trees and not for vegetables or fruits that are eaten raw. Additionally, greywater should be applied directly to the soil and not sprayed on plants to avoid contact with the foliage. This reduces the risk of contamination and ensures that the greywater is absorbed into the ground where it can be filtered naturally.

It's also important to use biodegradable, non-toxic soaps and detergents in your home. These products break down more easily in the environment and reduce the risk of chemical buildup in your garden soil, which could harm plant health over time. Making these small changes in your household product choices can have a big impact on the safety and effectiveness of your greywater recycling efforts.

Simple System Design

Setting up a basic home greywater system can be easy and affordable. The simplest systems usually involve altering the plumbing of a washing machine to divert the water to your garden instead of the sewer. This setup generally includes a three-way valve that allows you to switch between sending the water to the sewer during the winter (when you might not need the water for irrigation) or directing it to your garden during the growing season.

Here's a basic outline for setting up a simple greywater system from your washing machine:

1. **Install a Three-Way Valve:**
    - This valve is installed on the washing machine's discharge hose and allows you to control where the water goes.
    - Position it so you can easily access it to change the flow direction.
2. **Create a Filtration System:**
    - Attach a filter to the hose leading to your garden to remove lint and other particles. This can be as simple as a mesh bag or a commercially available greywater filter.
3. **Lay Out Irrigation Lines:**
    - Use standard irrigation hose or PVC pipes to distribute the water throughout your garden.
    - Ensure the pipes are laid out below the soil surface to reduce evaporation and contact with plant leaves.

4. **Use Mulch Basins:**

    - Direct the greywater to mulch basins around your plants. These are simply areas where mulch is piled up to distribute the water more evenly and prevent pooling, which can lead to root rot.

This setup provides an effective way to reuse a significant amount of water without a large investment in equipment or major alterations to your home's plumbing system.

Maintenance and Monitoring

Maintaining a greywater system is essential for its efficiency and longevity. Regular checks and cleaning of filters are crucial to prevent clogging and ensure smooth operation. It's also important to monitor soil conditions in areas where greywater is used, as changes in soil pH or salinity levels can indicate an accumulation of salts or other substances. If you notice plants showing signs of stress, such as yellowing leaves or stunted growth, it may be necessary to adjust your greywater usage or revisit your choice of detergents and soaps.

Keeping a log of system performance and any maintenance activities can help you track the system's health and troubleshoot any issues that arise. This record-keeping can be invaluable, especially if you need to demonstrate compliance with local regulations or want to assess the system's impact on your water usage and garden health over time. Be sure to contact your local city website or water authority for details of what is required.

Incorporating a greywater system into your gardening practices is a proactive step towards a more sustainable and water-wise lifestyle. As you refine your system and integrate it into your daily routine, you'll not only see the benefits in your flourishing garden but also feel a deeper connection to the cycles of use and reuse that define sustainable living. This project is more than just a technical setup; it's a commitment to stewarding resources wisely, ensuring that every drop of water serves a purpose in nurturing the world around you.

7.8 Crafting a Solar-Powered Water Feature

Installing a water feature in a garden located in an arid region, due to the high rates of evaporation, might seem counterintuitive at first. However, there are several reasons why this can be beneficial:

1. **Microclimate Creation:** A water feature can help create a microclimate within its immediate surroundings. The presence of water can cool the air and increase local humidity levels. This can be particularly refreshing in a hot, dry climate and can benefit the plants around the water feature by reducing the heat stress on them.

2. **Aesthetic Appeal:** They add a unique aesthetic element to a garden. The sight and sound of water can transform a landscape, making it more tranquil and appealing. In arid regions, where lush greenery might be harder to sustain, a water feature can serve as a focal point that adds visual interest and a sense of oasis.

3. **Attracts Wildlife:** A water feature can attract birds, insects, and other wildlife to a garden, promoting biodiversity. In arid regions, where water is scarce, a water feature can provide a vital water source for wildlife and contribute to conservation efforts.

4. **Thermal Regulation:** They can help regulate temperature. During the hottest parts of the day, the evaporation of water from the feature can cool the air around it. This can make outdoor spaces more comfortable and reduce the need for artificial cooling.

5. **Noise Masking:** In urban areas, water features can help mask traffic noise or other urban sounds with the soothing sounds of flowing or trickling water, creating a peaceful retreat.

To mitigate the high rates of evaporation, strategies can be employed:

Size and Depth: Deeper water features with less surface area relative to volume evaporate slower than shallow ones.

Shade: Placing water features in shaded areas or designing them with overhead covers can reduce evaporation by shielding the water from direct sunlight.

Water Conservation Technology: Using recirculating pumps ensures that water is reused efficiently, and incorporating sensors can maintain water levels efficiently to minimize waste.

Overall, the inclusion of a water feature in an arid garden can be justified by its multiple benefits, which outweigh the challenge of higher evaporation rates, especially if measures are taken to mitigate water loss.

DIY Project Overview

Creating a solar-powered water feature can be a rewarding weekend project that doesn't require extensive technical knowledge or a large budget. Start by choosing a location in your garden that receives ample sunlight; this is crucial as the solar panels need direct exposure to the sun for maximum efficiency. Once you've selected the perfect spot, the next step involves choosing the type of water feature. A simple fountain or a small waterfall can both be effective. The sound of a waterfall can be particularly soothing and adds a naturalistic element to your garden, while a fountain might be preferred for its decorative patterns of water spray.

To build a solar-powered fountain, you will need a solar panel, a submersible pump, and a basin to hold the water. There are many kits available that provide all the necessary components, parts, and installation instructions. First, secure the solar panel near the fountain; it should be close enough to connect to the pump but positioned to capture optimal sunlight. Attach the pump to the solar panel using the connectors provided—these should be straightforward plug-in connections. Place the pump in the basin, fill it with water, and ensure the fountain's spout or the waterfall's cascade is positioned to circulate the water effectively. Test the setup to ensure everything is functioning correctly, making adjustments to the solar panel's angle or the water flow as necessary.

Material Selection

Choosing the right materials for your solar-powered water feature is key to its success and longevity. For the basin and any structural elements, durable materials that can withstand the elements are essential. Natural stone or high-quality resin are popular choices, offering both durability and aesthetic appeal. For the solar panel and pump, opt for products specifically designed for outdoor use, with adequate weatherproofing and durability.

When selecting your materials, also consider the overall design and how it complements your garden's theme. A rustic, naturalistic garden might benefit from a stone basin and waterfall, while a sleek, geometric fountain might suit a more modern space. Whatever your choice, ensure it integrates well with the existing garden elements to create a cohesive and inviting outdoor space.

Wildlife Attraction

One of the most delightful benefits of having a water feature in your garden is its ability to attract wildlife. Birds, beneficial insects, and even small mammals are drawn to the sound and sight of water, especially in urban or dry areas where natural water sources may be scarce. The water feature acts as a beacon for these creatures, providing them with a necessary resource and enhancing the ecological dynamics of your garden. Birds, in particular, not only use the feature for drinking and bathing but also help control pests by eating insects. Additionally, the presence of water can encourage pollinators such as bees and butterflies, contributing to the health of your flowering plants.

Incorporating features such as shallow edges on your basin or flat stones within the water can make the feature more accessible to smaller wildlife, ensuring that it is safe and beneficial for a variety of creatures. By thoughtfully designing your water feature, you not only create a stunning visual and auditory focal point but also contribute to local biodiversity, supporting nature's delicate balance right in your own backyard.

7.9 Sourcing Materials for DIY Projects on a Budget

When embarking on DIY projects, sourcing materials on a budget can be both a challenge and an opportunity to get creative. The key to economical DIY is knowing where to look for materials that can be repurposed, recycled, or bought at a significant discount.

Firstly, second-hand stores, thrift shops, and garage sales are treasure troves for DIY enthusiasts. Items like old furniture, frames, jars, and tools can often be found at a fraction of the cost of new ones. These venues are especially useful for finding unique or vintage items that can add a distinctive touch to your projects. It's worth visiting these places regularly as their inventory changes frequently.

Another great resource is online marketplaces. Websites like Craigslist, Freecycle, or local Facebook groups can be excellent for finding free or cheap materials. People often give away leftovers from their own projects, like wood, tiles, or paint, which can be perfect for small projects. Always ensure to meet in public spaces or go with a friend when picking up items from someone you met online.

Construction and renovation sites can also be sources of free materials. With permission, you can often collect scrap materials that would otherwise be disposed of. This can include wood, bricks, and other building materials that are ideal for larger projects. Just be sure to get permission and prioritize safety, both yours and the site's.

Finally, consider repurposing materials you already have at home. Old clothes can be turned into rags or quilts, broken dishes can become mosaic tiles, and jars can be used for storage solutions. This approach saves money and reduces waste. Sourcing materials for DIY projects on a budget requires a mix of creativity, diligence, and a willingness to look in less obvious places. By combining these strategies, you can find affordable materials that will help bring your DIY projects to life without breaking the bank.

As this chapter concludes, reflect on the transformative power of these DIY projects, from the ancient practice of using ollas to the modern implementation of solar-powered water features. Each project not only enhances the beauty and functionality of your garden but also deepens your connection to the environment through sustainable practices. Armed with these insights and inspirations, you are well-prepared to continue enhancing your dry garden, making each choice a step toward a more sustainable and vibrant outdoor space.

Chapter 8

Overcoming Common Challenges

Imagine your dry garden as a serene oasis, a harmonious blend of form and function where every plant thrives with minimal water. Yet, even in the most meticulously planned gardens, nature has a way of presenting challenges, particularly pests and diseases that can thrive under the same low-water conditions designed to sustain your drought-resistant plants. Understanding these challenges, preemptively planning to avoid them, and knowing how to address them when they arise can ensure that your garden remains a thriving refuge in an arid climate.

8.1 Handling Pests and Diseases in Low-Water Gardens

Identifying Common Pests and Diseases

The first step in protecting your garden is to recognize the enemy. In low-water conditions, certain pests and diseases find a favorable environment, exploiting the stressed plants struggling under the harsh conditions. Aphids, spider mites, and whiteflies are among the common culprits drawn to weakened plants. These pests typically cause damage by sucking sap from the leaves, leading to yellowing, stunted growth, or even plant death if not controlled. On the disease front, fungal infections like powdery mildew thrive in warm, dry climates, coating plant leaves with telltale white dust.

It's crucial to regularly inspect your garden, paying close attention to the undersides of leaves and the points where leaves attach to stems—common hiding spots for pests. Early signs of disease may include discolored spots on leaves, wilting, or abnormal growth patterns. By catching these signs early, you can manage them before they become a severe problem.

Preventive Measures

Prevention is always your first and best line of defense. Choosing the right plants is paramount; native or drought-tolerant species are naturally more resistant to pests and diseases prevalent in your area.

Good garden hygiene plays a critical role in preventing disease. Remove any fallen debris and spent foliage regularly to minimize the risk of fungal spores taking hold. Proper spacing between plants ensures good air circulation, which helps keep the foliage dry and less hospitable to pests and diseases. Additionally, watering practices can have a significant impact—water early in the morning to allow any moisture on the leaves to dry during the day, reducing the risk of fungal diseases.

Eco-Friendly Management

When prevention isn't enough, eco-friendly management techniques can help control pests and diseases without resorting to harsh chemicals that could damage the environment and the beneficial organisms in your garden. For pests, neem oil, insecticidal soaps, and horticultural oils are effective treatments that target pests without harming plants or beneficial insects. For diseases, products containing sulfur or copper can be effective, although they should be used as a last resort due to their potential impact on soil health and microbial life.

Case Studies

Consider the experience of a gardener in Tucson, Arizona, who successfully managed whiteflies using a combination of reflective mulch to deter the pests and introducing ladybugs, beneficial insects that prey on whiteflies. This approach not only controlled the whitefly population but also enhanced the biodiversity of the garden, contributing to its health and resilience.

Another example comes from a community garden in Santa Fe, New Mexico, where gardeners faced recurring bouts of powdery mildew. By switching to sulfur-based fungicides and implementing rigorous sanitation practices, they were able to protect their plants without resorting to more toxic chemicals. These changes not only improved the health of the garden but also aligned with the community's commitment to sustainable gardening practices.

Understanding the dynamics of your garden's ecosystem and being proactive in managing pests and diseases can significantly enhance the resilience of your dry garden. By focusing on prevention, opting for eco-friendly solutions, and learning from the experiences of others, you can maintain a healthy garden that conserves water and thrives in an arid climate.

8.2 Scaling Dry Gardening Practices for Large Areas

When you think about dry gardening, the image that often springs to mind might be small, meticulously managed plots where every drop of water is accounted for. But what happens when you scale these principles up to larger landscapes like public parks, agricultural lands, or even golf courses? The challenges multiply, certainly, but so do the opportunities to make a significant environmental impact. Expanding dry gardening practices to large areas involves a strategic blend of design, management, and maintenance that adapts to the expansive scale while still adhering to the core principles of water conservation and sustainability.

One of the primary challenges in scaling up is the design and implementation of efficient irrigation systems that can service large areas without wasting water. Traditional irrigation methods, such as sprinkler systems, are often not suitable due to their high water use and significant evaporation rates. Instead, drip irrigation emerges as a frontrunner in efficiency. For large landscapes, the design of such systems needs to be meticulously planned to ensure coverage and efficiency. This includes creating zones based on plant water needs, which allows for targeted irrigation schedules that match the specific water requirements of different plant areas.

The use of technology can greatly enhance irrigation efficiency in large-scale dry gardens. Automated irrigation systems equipped with moisture sensors can adjust watering schedules based on real-time soil moisture levels, ensuring that plants receive water only when necessary. These systems can be monitored and controlled remotely, providing a convenient solution for managing large gardens with precision. By integrating these technological tools, large-scale gardening projects can achieve a balance between maintaining healthy plant growth and minimizing water use.

Maintaining a large dry garden also presents unique challenges, particularly in terms of labor and resource allocation. The sheer size of these projects demands a well-coordinated effort to manage tasks such as planting, weeding, pruning, irrigation, and monitoring. This is where efficient garden management practices come into play. Utilizing a zone-based approach not only helps in irrigation management but also simplifies maintenance tasks. Each zone can utilize multiple moisture monitors and be assigned specific maintenance schedules based on the plants' needs, which streamlines operations and ensures that all areas of the garden receive the attention they require without unnecessary labor expenditure.

Moreover, the selection of plants for large dry gardens requires careful consideration. The key is to choose species that are not only drought-resistant but also capable of thriving in the local climate and soil conditions without extensive maintenance. Native plants are often ideal as they are adapted to the local environment and support local biodiversity. Incorporating a diverse mix of trees, shrubs, and ground covers can help create a resilient plant community that is more resistant to pests, diseases, and climatic fluctuations.

Case Studies

The Rio Salado Habitat Restoration Area in Phoenix, Arizona, is a remarkable example of urban environmental restoration. This 600-acre site along the Salt River was transformed from a neglected, debris-laden riverbed into a thriving riparian ecosystem. The project was initiated to restore the native wetland and riverside habitats that were historically prevalent along the river before urban development and water diversion altered the landscape.

The area has been replanted with a variety of native species that are adapted to the local climate. This helps to conserve water and provides a more suitable environment for local wildlife.

The restoration has created a sanctuary for over 200 species of birds and other wildlife, making it a popular spot for bird watching and nature photography. The site features several miles of trails, both paved and natural, that allow for hiking, cycling, and horseback riding, catering to a wide range of outdoor enthusiasts. The Rio Salado Audubon Center located within the area provides educational programs that focus on conservation and the environment, offering a hands-on learning experience about the importance of habitat restoration.

The restoration efforts at Rio Salado are part of a broader initiative to manage natural resources sustainably and enhance the quality of life in urban areas by reintegrating nature into the cityscape. This project benefits the environment by restoring an essential habitat and serves as a recreational and educational resource for the Phoenix community.

A notable project in Silicon Valley that underwent a similar transformation is the Microsoft Silicon Valley Campus in Mountain View, California. This campus, originally featuring traditional landscaping, was redesigned using principles of xeriscaping and sustainable landscaping to emphasize sustainability and ecological restoration. The project included the installation of drought-tolerant grasses, a variety of succulents, and permeable paving materials to create an aesthetically pleasing yet water-efficient environment. The irrigation system was upgraded with moisture sensors and a weather-based control system, significantly reducing the campus's water consumption by over 50%. This project reflects Microsoft's commitment to creating a green, sustainable workplace that aligns with the company's overall sustainability goals.

These examples underscore the scalability of dry gardening principles when applied with thoughtful planning and innovative technology. By addressing the unique challenges of large-scale landscapes, it is possible to create extensive green spaces that are both beautiful and environmentally sustainable. Whether it's a public park, an agricultural field, or a corporate garden, the principles of dry gardening provide a roadmap for managing these spaces in ways that conserve water, support ecosystems, and create sustainable, functional landscapes for everyone to enjoy.

8.3 Maintaining Aesthetic Appeal in the Dry Season

When the rains retreat and the land dries, maintaining the vibrancy and allure of your garden can seem like a daunting task. The dry season often strips the garden of its lush greenery, leaving behind a more subdued palette that might feel less enchanting. However, with thoughtful planning and strategic design, your garden can continue to captivate the senses and provide a serene retreat no matter the season.

The secret to a garden that maintains its appeal year-round lies in understanding and creatively responding to the cyclical changes in plant life and garden conditions. One effective strategy is to design for year-round interest by incorporating structural elements that remain constant regardless of the season. Hardscaping features such as rock formations, sculptures, pathways, and benches can provide visual interest and form the backbone of the garden landscape. These elements not only add texture and contrast but also help define the space, guiding the eye and moving the visitor through the garden in a deliberate, thoughtful way.

Water features like fountains or birdbaths can play a pivotal role in enlivening a dry-season garden. They introduce movement and sound, which bring dynamic elements to the garden that are particularly appealing when plant life is dormant. Additionally, water features attract wildlife such as birds and butterflies, adding life and activity to your garden. Even in their simplest forms, these features can transform a static scene into a lively tableau, enriching the garden's sensory appeal.

Incorporating evergreens into your landscape is another effective strategy for ensuring continuous visual interest. Evergreens do not lose their leaves and remain vibrant throughout the year, providing a constant hue of green that can be both calming and uplifting. They serve as excellent focal points or backdrops against which the seasonal cycles of flowering plants can unfold. The strategic placement of evergreens can help mask the less visually appealing phases of other plants that may lose their leaves or flowers during the dry season.

Color and Texture During the Dry Season

To further enhance your garden's aesthetic appeal during the dry season, consider the use of plants and garden elements that add bursts of color and varied textures. Plants with interesting visual characteristics, such as desert willow or mesquite trees, can provide visual interest even without foliage. Similarly, ornamental grasses, with their rich textures and movement, can become standout features in a less verdant landscape.

Choosing plants with vibrant berries or persistent fruits can also add splashes of color to the garden. For example, wolfberry produces bright red or orange berries, and prickly pear can offer purple or reddish fruits that linger on branches. These plants not only enliven the garden visually but can also serve as important food sources for local wildlife during the colder months, thus supporting biodiversity.

Succulents and cacti are particularly suited to dry conditions and can be used to great effect in adding structure and interest to the garden. Their varied forms and textures—from the rosettes of Echeveria to the towering columns of cereus cacti—make them excellent choices for creating focal points in the landscape. Many succulents also flower in response to the onset of the dry season, providing unexpected blooms when other plants are dormant.

Success Stories

Let's draw inspiration from the Mediterranean Garden at the Denver Botanic Gardens, an exemplary model of aesthetic resilience in the face of dry conditions. This garden beautifully integrates drought-tolerant plants with year-round appeal, featuring a tapestry of lavender, sage, and thyme interspersed with striking agave and aloe. The garden's design emphasizes texture and form, using gravel paths and stone walls to enhance the natural beauty of the plantings. Despite Denver's dry summers, the garden remains a lush and inviting space, demonstrating the potential to create a visually appealing landscape that is both sustainable and beautiful.

Another inspiring example comes from a private residence in Santa Fe, New Mexico, where the homeowner transformed a traditional lawn into a vibrant xeriscape garden. This garden features a mix of native grasses, perennials, and shrubs that provide color and texture throughout the year. Strategic placements of sculptures and a small, recirculating stream add artistic touches that make the garden a delightful discovery in every season. The use of mulches and ground covers in various colors and textures further enriches the visual palette, ensuring that the garden is as appealing in the dry months as it is in the more verdant seasons.

By embracing these strategies and learning from successful examples, you can ensure that your garden remains an enchanting sanctuary throughout the dry season. It's about viewing the garden not just as a collection of plants but as a holistic composition of colors, textures, and structures that together create a harmonious and engaging space.

8.4 Adapting Water-Saving Techniques in Humid Climates

In regions where humidity hangs heavy in the air and rain seems more a rule than an exception, the principles of dry gardening can feel paradoxical, if not downright impractical. Yet, it's precisely in such environments that the wisdom of water conservation and sustainable gardening can shine, provided they are thoughtfully adapted to the local context. One might wonder how techniques forged in the aridity of deserts could apply to a milieu where moisture is abundant. The challenge lies not in the surplus of water, but in managing it wisely to prevent issues like root rot, fungal diseases, and nutrient leaching which are prevalent in humid climates.

Adapting dry gardening techniques to humid areas starts with a fundamental shift in how we view and manage water. Even in water-abundant settings, conserving water remains crucial, as excessive reliance on irrigation can lead to inefficient use of this resource and contribute to water pollution due to runoff. The first step is improving soil drainage. This can be achieved by amending the soil with materials like coarse sand, perlite, and organic compost, which enhance its structure, prevent compaction, and facilitate better water passage. Raised beds are another effective strategy; they offer excellent drainage and can help prevent the waterlogging that is so detrimental in humid climates. They also warm up faster in the spring, extending the growing season for many plants.

When it comes to irrigation, the principles of dry gardening advocate for deep, infrequent watering that encourages plants to develop deep root systems, making them more drought-resistant and stable. This technique is beneficial even in humid climates as it helps plants withstand periods without rain and reduces surface water wastage. Utilizing drip irrigation systems can be particularly effective, as they deliver water directly to the root zone where it's needed most, minimizing evaporation and runoff. These systems can be adjusted to the rainfall patterns of humid regions, ensuring that plants receive just the right amount of water without excess.

Selecting appropriate plants plays a pivotal role in the success of dry gardens in humid regions. It involves choosing species that are not only drought-tolerant but also capable of thriving in high-moisture environments. Plants such as ferns, which naturally inhabit moist, shady areas, can be excellent choices. However, it's also possible to grow more typically arid-region plants if they are given adequate drainage as discussed earlier. For instance, lavender, which is generally associated with dry Mediterranean climates, can thrive if planted in raised beds with well-drained soil and full sun exposure in a humid climate. This adaptability underscores the importance of understanding plant needs and local environmental conditions in garden planning.

Several other plants that thrive in regions with higher natural rainfall but are also drought-resistant include:

1. **Astilbe (Astilbe spp.):** These are popular in shady gardens due to their feathery, plume-like flowers and ability to thrive in both moist soils and periods of drought once established. They prefer partial to full shade and make a colorful addition with their red, pink, white, or purple blooms.

2. **Bee Balm (*Monarda didyma*):** Known for its aromatic leaves and striking, spiky flowers, bee balm is excellent for attracting pollinators like bees and hummingbirds. It can handle damp conditions but also copes quite well during dry spells, especially when mulched to retain soil moisture.

3. **Japanese Iris (*Iris ensata*):** These irises excel in wetter soils and are often used in water garden settings. They are also quite drought-tolerant once established, making them versatile for varying moisture conditions. They produce large, elegant flowers that can be a highlight in any garden.

4. **Swamp Milkweed (*Asclepias incarnata*):** As its name suggests, this plant is accustomed to wet environments but is surprisingly drought-tolerant as well. It's especially valuable for attracting butterflies, including monarchs, and offers vibrant pink and light mauve flowers.

5. **Siberian Iris (*Iris sibirica*):** This hardy iris thrives in moist conditions and is also drought-tolerant once established. It features beautiful, grass-like foliage and striking blue or purple flowers, adding an elegant touch to the garden.

6. **Joe-Pye Weed (*Eutrochium purpureum*):** Despite its weedy name, Joe-Pye weed is a stately plant that can reach several feet in height and sports fluffy, mauve-pink blooms. It's naturally found in moist areas but can withstand periods of drought, making it ideal for gardens in humid regions with occasional dry spells.

These plants not only manage well in fluctuating moisture levels but also contribute to the biodiversity of the garden by providing habitat and food for various wildlife species.

Real-World Adaptations

Let's take the example of a garden in coastal Florida, a region known for its high humidity and frequent rainfall. Here, a gardener named Elena successfully adapted dry gardening principles to create a thriving landscape around her home. She replaced her lawn with native ground covers like coontie and sunshine mimosa, which are well-adapted to local conditions and require little to no supplemental watering. She also incorporated rain gardens in areas where water tended to collect, using moisture-loving plants such as swamp milkweed and horsetail reed. These rain gardens help to filter and slowly release the collected water, preventing runoff and making efficient use of the abundant rainfall.

Another example comes from the Pacific Northwest, known for its wet, cool winters and dry summers. A community garden in Portland implemented a system of swales—shallow, vegetated trenches that capture and direct runoff. These swales are planted with drought-resistant plants that can withstand the dry summer months, such as Oregon grape and Douglas aster. During the rainy season, the swales manage excess water effectively, reducing the burden on the city's drainage system and enhancing the local ecosystem's resilience.

These examples illustrate that with thoughtful modification, the principles of dry gardening can be effectively applied even in humid climates. By focusing on soil health, appropriate plant selection, and innovative water management techniques, gardeners in these regions can enjoy the benefits of sustainable, water-efficient gardening. This approach not only conserves an important natural resource but also creates a garden that is both beautiful and resilient, capable of adapting to the unique challenges and opportunities of its environment.

8.5 Adapting to Disabilities

Dry gardening, which focuses on using drought-resistant plants and water-conserving techniques, can be particularly adaptable and beneficial for gardeners with various disabilities. Here are some ways it can be tailored:

1. **Raised Beds and Containers:** These can be used to bring the garden to a more accessible height for individuals in wheelchairs or those who have difficulty bending down. Raised beds can also be designed with wide, sturdy edges for seating support.

2. **Drip Irrigation Systems:** These systems minimize the need for manual watering and can be automated to reduce physical strain. Drip irrigation delivers water directly to the base of the plant, which is efficient and reduces the maintenance work.

3. **Pathway Accessibility:** Ensuring that pathways are wide and stable can make garden areas accessible to those with mobility aids, such as wheelchairs and walkers. Materials like compacted fine gravel or smooth concrete can provide stable surfaces.

4. **Tool Adaptation:** Ergonomically designed tools can make gardening more accessible. Tools with extended handles or enhanced grips can help individuals with limited mobility or strength.

5. **Sensory Gardens:** Incorporating plants with different textures, scents, and colors can enhance the gardening experience for individuals with visual impairments or cognitive disabilities. Labeling plants in Braille or using high-contrast colors for signage can also help.

6. **Low Maintenance Plants:** Choosing plants that require less frequent care, such as succulents and other drought-resistant varieties, can make gardening more manageable for those with limited energy or physical limitations.

7. **Educational Resources:** Providing accessible formats for gardening guides, such as audio books, large print, or videos with closed captioning and sign language, can help make information more accessible to everyone.

By implementing these adaptations, dry gardening can be a fulfilling activity that is inclusive of gardeners with diverse abilities.

Residents at a Phoenix, Arizona retirement community have created a state-of-the-art, accessible, raised bed garden to ensure that everyone, including those with disabilities, have convenient access to relax and enjoy the beautiful surroundings. Key features include ADA-compliant ramps with railings, 3-foot wide pathways with 5-foot turnaround spaces, wheelchair accessible beds, and 29-inch high raised beds for easy access.

Paths have contrasting colors for visibility, and the surface is smooth and compacted, making it suitable for walkers, wheelchairs, and electric scooters. In addition to an automatic irrigation system, supplemental watering is made easy with large-handled shutoff valves, conveniently placed hoses, and one-touch spray nozzles. Safety is prioritized with grab bars at the tool and storage shed entrance and armrests on chairs and benches.

This garden is designed to be a welcoming space for all community residents, allowing everyone to experience the joy of gardening.

Sources for disability-adapted tools

There are several excellent sources for ergonomically designed gardening tools that cater to people with disabilities:

1. **Amazon:** Offers a wide range of ergonomic gardening tools like the Peta Easi-Grip Long Reach Garden Tools, which are ideal for those who prefer to garden from a seated position or have limited flexibility. These tools are designed with non-slip handles and soft grips to accommodate gardeners with arthritis.

2. **Arthritis Supplies:** Features tools like the Garden Kneeler and Seat, which helps reduce strain on the knees and also provides a seat for breaks. This site also offers tools with Easi-Grip handles, which are designed to keep the wrist in a natural position, reducing strain while gardening.

3. **Ability Superstore:** This site offers a range of gardening aids designed for individuals with limited mobility, including long-handled tools and ergonomic handles that can be added to existing tools. Their selection is geared towards making gardening more accessible and enjoyable.

4. **Essential Aids:** Provides a variety of gardening tools specifically adapted for the elderly and disabled, such as long reach trowels, forks, and ergonomic garden tools with arm support cuffs. They also have special

offers and VAT relief on many items, which can make these tools more affordable.

5. **Best Mobility Aids:** Another great source for tools like easy-grip tools and long-handled equipment, which are particularly useful for those who cannot bend or stoop easily. They also offer Bionic Relief Gardening Gloves designed by an orthopedic hand surgeon for arthritis sufferers, providing extra support around the wrists.

Each of these retailers provides tools that are designed to make gardening more accessible and enjoyable for people with various disabilities, emphasizing comfort, ease of use, and reduced physical strain.

As we wrap up this chapter, we transition into exploring the role of community engagement in dry gardening, where the principles of sustainability and resource management are shared and promoted, further enriching our connection to the environment and each other.

Chapter 9

Case Studies: Success Stories in Dry Gardening

9.1 Urban Apartments: Maximizing Minimal Space

Imagine transforming a concrete balcony or part of an interior room into an oasis of greenery, where every inch is a testament to the ingenuity and possibility of urban gardening. For many city dwellers, their apartment is not just their living space but a canvas for creativity and sustainability. This section celebrates those urban gardeners who have embraced the challenge of limited space to create lush, water-efficient gardens, flourishing high above the bustling city streets.

Innovative Container Gardening

In the heart of the city, where ground is scarce, the skies are the limit for gardening. Urban gardeners have turned to containers and vertical gardening structures to maximize their green spaces. By using stackable planters, hanging baskets, and even repurposed items like old pallets and shoe organizers, they have defied spatial limitations, creating vertically sprawling gardens. These innovative approaches not only optimize the available space but also add a vibrant layer of beauty and privacy to their urban abodes.

The utility of container gardening in such settings extends beyond aesthetics. It allows for endless rearrangement possibilities, and precise control over soil quality and moisture levels, crucial in urban environments where natural earth is inaccessible. Containers can be filled with custom soil blends tailored to the specific needs of each plant, ensuring that even in a concrete jungle, a diverse range of plants can thrive. This controlled environment significantly reduces the risk of soil-borne diseases and pests, common in more traditional garden settings, and offers a perfect solution for organic gardening enthusiasts who wish to avoid pesticides.

Selection of Suitable Plants

Choosing the right plants is pivotal for the success of a balcony garden, especially one that aims to conserve water. Drought-resistant plants such as sedums, lavender, and echeverias not only survive but thrive in containers with minimal watering. Herbs like rosemary, thyme, and sage, which are accustomed to the rocky, dry soils of the Mediterranean, offer practical, culinary additions to the garden while requiring little water. These plants are not only water-wise but also resilient, able to withstand the variable conditions typical of an elevated urban setting, from scorching sun to gusty winds.

For those looking to add a splash of color, flowering plants like marigolds and petunias can also be included. While slightly more demanding in terms of water, they can still be part of a water-efficient garden if paired with appropriate watering techniques and soil mixtures that enhance moisture retention, such as those containing coconut coir or peat moss.

Water-Saving Tips and Tricks

Efficient water use is a cornerstone of successful dry gardening on balconies and in apartments. One key strategy is the use of ollas. More information on this ancient watering method is provided in chapters 3 and 7.

Self-watering containers offer another effective solution, especially for the busy urbanite. These containers have reservoirs at the bottom that allow plants to absorb moisture as needed, reducing the frequency and volume of watering. By using these methods, balcony gardeners can reduce their water usage significantly.

Success Stories

Let's look at a couple of examples that illustrate how individuals in different urban settings have successfully adapted dry gardening techniques to their unique living environments, promoting sustainability and a connection to nature despite space limitations.

Our first story takes place in a compact apartment in Las Vegas, Nevada. Jamie is a graphic designer who works from home. She transformed one corner of her small living room into a dry garden oasis. Given limited space and lack of natural soil, she opted for a collection of large clay pots and vertical planters. The pots are filled with a sandy, well-draining potting mix, ideal for dry gardening. Her selection of plants includes succulents; Echeveria, jade plants, and aloe vera, which thrive indoors with minimal water. Jamie also incorporated a small desert rose and several cacti varieties to add diversity and texture.

An innovative drip irrigation system was set up to provide minimal, consistent moisture directly to the roots with almost no waste. Jamie used a timer on the drip system to water the plants sparingly, that mimicks their natural arid conditions.

Jamie enjoys the low-maintenance aspect of her dry garden, which adds a touch of

greenery to her apartment without requiring frequent care. The primary challenge was adjusting lighting conditions, so she installed LED grow lights to supplement natural sunlight.

Our next case study takes us to a high-rise apartment in Tucson, Arizona. Carlos, a retired school teacher and amateur botanist, utilized several large pots made from recycled containers on his balcony, which faces south, receiving abundant sunlight most of the day. His garden boasts a variety of native desert plants such as small yucca, sagebrush, and agave. To add color and attract pollinators, Carlos planted Indian blanket and desert marigold. As an amateur chef, Carlos also included some thyme, sage, and oregano in his garden and next year is planning to add rosemary and lavender for a refreshing herbal tea.

Carlos designed a gravity-fed watering system that collects rainwater in a barrel. A simple hose with drip nozzles allows for slow, deep watering that encourages root growth and minimizes evaporation in the hot, dry climate. When needed, Carlos tops off the water in the barrel to provide a steady supply.

The balcony garden serves as a practical demonstration of water-wise gardening in an urban setting. Carlos enjoys the beauty and biodiversity of his garden, which also serves educational purposes when he hosts local student groups. The main challenge he faces is protecting his plants from high winds and adjusting the garden layout to optimize exposure to the sun.

These stories underscore the adaptability and creativity of urban gardeners who manage to cultivate stunning green spaces in minimal areas. They prove that, with the right techniques and choices, even a corner of a living room or the smallest balcony can be turned into a productive and environmentally friendly garden. As we move through this chapter, let these stories inspire you, showing that no space is too small to have a significant impact.

9.2 Suburban Dry Gardens: Transforming Lawns into Oases

In the suburban sprawl where traditional grassy lawns once ruled, a quiet revolution is taking place. Homeowners are increasingly turning away from the lush, high-maintenance turf to embrace the beauty and practicality of drought-tolerant landscapes. This shift is not just about aesthetics; it's a conscious decision to foster sustainability in personal outdoor spaces. By transforming their lawns into dry gardens, these suburban pioneers are reducing their water usage dramatically, cutting down on maintenance, and creating serene landscapes that resonate with the natural environment.

One of the most compelling reasons for this shift is the significant reduction in water consumption. Traditional lawns require an immense amount of water, especially during the warmer months. By replacing these with native, drought-resistant plants, homeowners can minimize their irrigation needs substantially. Additionally, these dry gardens require less frequent mowing, fertilizing, and pest control, further reducing the garden's environmental impact and the homeowner's workload.

Designing a dry garden, however, involves more than just planting drought-tolerant species; it requires a thoughtful approach to landscape design that harmonizes with the home's architecture and the surrounding neighborhood. The key elements of successful dry garden design include the use of native plants that are adapted to the local climate and soil conditions, which ensures better growth and resilience.

Mulching is another critical aspect of the conversion of lawn areas into dry gardens. It helps retain soil moisture, suppress weeds, and provide a clean and finished look to the garden beds. Strategic placement of hardscaping elements like stones, gravel, or wood chips can add contrast and structure to the garden, creating pathways and defining areas. These elements help reduce the soil area by cutting down on water use and serving as heat-reflecting surfaces that reduce temperatures around the plants.

The broader impact of transforming a lawn into a dry garden extends well into the community. Such gardens can serve as inspiration for neighbors, sparking a broader change towards more sustainable landscaping practices in the area. They also contribute to local biodiversity by providing habitats for native wildlife, including birds, bees, and butterflies.

Featured Success Stories

Take, for example, the story of the Thomson family in Phoenix, Arizona, known for its arid climate. They replaced their thirsty Bermuda grass lawn with a vibrant assortment of native desert plants like blue agave, red yucca, and various succulents. Not only has their water bill decreased by 40%, but their garden has also become a local attraction, inspiring several neighbors to reconsider their own landscaping choices.

Another impactful story comes from the suburbs of Los Angeles, where the Garcias turned their backyard into a sustainable retreat. By incorporating elements such as a small rock garden, a dry creek bed, and a variety of California-native plants, they created a stunning landscape that requires minimal upkeep and water. Their garden has not only reduced their household's ecological footprint but has also become a haven for local birds and beneficial insects, enhancing local biodiversity.

These stories are just a couple of examples of how individual choices in garden design can lead to substantial community and environmental benefits. They illustrate the power of personal action in fostering larger changes. As more homeowners adopt these practices, the cumulative effect can lead to significant ecological benefits, including reduced urban heat island effect and enhanced local air and water quality.

This wave of change towards dry gardening in suburban landscapes reflects a growing recognition of the need for sustainable living practices. It demonstrates a shift in how we interact with our environment, opting for harmony rather than control, and sustainability over short-term convenience. As this movement grows, it paves the way for a future where dry gardens are a standard feature of suburban development.

9.3 Commercial Properties Embracing Dry Gardening Practices

In today's eco-conscious marketplace, the aesthetic and functional appeal of commercial properties can significantly influence business success. More and more, companies are recognizing that sustainable landscaping isn't just an environmental statement but a potent tool to attract like-minded customers and enhance their brand image. Across bustling city centers and sprawling retail complexes, an increasing number of commercial properties are integrating dry gardening principles into their landscapes.

The adoption of dry gardening practices in commercial settings involves replacing traditional, high-maintenance turf with drought-resistant plants and innovative water-saving designs. This transformation is often spearheaded by forward-thinking business owners and property managers who understand that long-term savings from reduced water and maintenance costs can significantly outweigh initial setup expenses. Replacing grassy areas with native plant gardens or decorative rock can drastically cut down on the need for irrigation and frequent upkeep. Moreover, these changes can often catch the eye of potential customers or tenants who value environmental sustainability.

The design and implementation of dry gardens in commercial properties can pose unique challenges, particularly due to the scale and public nature of these projects. Unlike private gardens, commercial landscapes must withstand heavier foot traffic and greater public scrutiny. Thus, the planning phase is critical and often involves collaboration with landscape architects who specialize in xeriscaping—the practice of designing landscapes to reduce or eliminate the need for supplemental irrigation.

Furthermore, the integration of hardscaping elements like pathways, patios, and retaining walls enhance the functionality of the space and reduce its water demand. These features help manage foot traffic and direct attention to focal points such as water features or sculptures. The choice of materials, such as recycled concrete or native stones, also plays a crucial role in reinforcing the property's green credentials.

Highlighting standout commercial properties that have successfully integrated dry gardening into their landscaping showcases the potential and versatility of these practices. Fashion Valley and Westfield UTC Malls in San Diego, California are two notable examples of large shopping centers which have revamped their traditional landscaping to incorporate native plants and sophisticated rainwater collection systems. Not only has this substantially reduced their water usage, but it has also become a key feature of their marketing strategies, attracting tenants and shoppers who prioritize sustainability.

Another example of an inspiring project is the Mueller Office Park in Austin, Texas, a corporate office complex which transformed its exterior grounds into a sustainable garden that doubles as a recreational area for employees and visitors. The garden features native Texas grasses and perennials that require minimal irrigation, shaded seating areas made from recycled materials, and a series of interconnected walkways that encourage walking meetings and outdoor activities. This initiative not only reduced the complex's operational costs but also enhanced employee satisfaction and retention.

These examples illustrate the environmental and economic impacts of adopting dry gardening practices as well as transforming commercial spaces into sustainable, functional, and appealing environments. As more businesses recognize these benefits, dry gardening is likely to continue gaining traction in the commercial sector, setting new standards for landscape design in business settings. As commercial properties continue to innovate in this area, they contribute to a larger movement towards sustainability, influencing industry standards and consumer expectations alike.

9.4 Community Gardens and Dry Gardening Education

Community gardens have become vibrant hubs of education and interaction, particularly in urban settings where green space is limited. These shared plots are not only about growing plants but also about cultivating community ties and spreading knowledge on sustainable practices. In the realm of dry gardening, community gardens serve as crucial educational platforms, providing hands-on learning experiences about water conservation and environmentally friendly gardening techniques. These gardens become outdoor classrooms where people of all ages and backgrounds can come together to learn, share, and grow.

In Chapter 10 we will explore methods of forming your own local group of dry gardeners and suggesting activities and ways of generating interest within your community.

Educational Outreach

The educational value of community gardens extends beyond simple plant cultivation. They offer a dynamic environment to teach and demonstrate the principles of dry gardening, such as proper soil amendment, the choice of drought-resistant plants, and efficient watering techniques. Workshops and seminars can be organized to cover a range of topics, from the basics of setting up a dry garden to more advanced topics like permaculture design principles. These sessions not only teach practical gardening skills but also help to instill a deeper understanding of ecological balance and the importance of resource conservation.

Moreover, community gardens often partner with local environmental groups, universities, and experts to provide a broad spectrum of learning opportunities. This form of outreach is particularly effective as it combines theoretical knowledge with practical, hands-on gardening experience, making the learning process both engaging and productive.

Collaborative Gardening Efforts

Community gardens are inherently collaborative spaces, and this is vividly reflected in their approach to water conservation and dry gardening. By pooling resources and knowledge, gardeners can implement more extensive and effective water-saving measures than they might individually. For example, a community garden can invest in a shared rainwater harvesting system or a solar-powered irrigation system, which might be too costly or complex for a single household to manage.

These collaborative efforts often lead to the development of innovative gardening techniques tailored to specific local conditions. Gardeners can experiment with different approaches to water management, plant selection, and garden design, learning from each other's successes and failures.

Engaging Diverse Communities

One of the most significant strengths of community gardens is their ability to engage a wide range of participants. From school children and families to retirees and immigrants, community gardens attract a diverse group of people, each bringing their unique perspectives and experiences to the garden.

To effectively engage these diverse groups, community gardens often employ a variety of outreach strategies. For children and schools, they might offer educational programs that align with school curriculums, providing a practical context for environmental science lessons. For adults, they might offer evening or weekend workshops that cover more advanced topics. Furthermore, community gardens can act as cultural bridges, incorporating traditional gardening techniques from different cultures into their practices, which not only enhances the garden's diversity but also helps in preserving and sharing cultural heritage.

Case Studies

An exemplary case is the Green Thumb Community Garden in San Jose, CA. What started as a barren lot has transformed into a lush garden that practices and teaches dry gardening techniques. Through a series of workshops and community involvement initiatives, the garden has educated hundreds of locals on creating sustainable, water-efficient green spaces. Their programs on rainwater harvesting have been particularly successful, leading to the installation of several new systems in homes around the community.

Another inspiring example comes from the Riverside Community Garden in New London, CT, which has made significant strides in engaging local schools. They have developed a curriculum-based program that integrates dry gardening practices into the science curriculum of nearby elementary schools. This program not only teaches children about the importance of water conservation but also gives them a sense of stewardship for their environment. The success of this initiative has sparked interest in other schools in the region, leading to a growing network of educational community gardens.

Through these community gardens, the principles of dry gardening are not just being practiced; they are being lived and taught to the next generation. These gardens stand as testaments to the power of collective effort and education in fostering environmental awareness and action, proving that even the smallest seed of knowledge can grow into substantial change.

9.5 Schools and Educational Programs Implementing Dry Gardens

In the evolving landscape of academic environments, schools are increasingly becoming fertile ground for integrating eco-conscious practices like dry gardening. To educators gardens serve as more than just aesthetic enhancements; they are dynamic classrooms offering students hands-on experiences in ecology, water conservation, and sustainable living. When students interact with a dry garden, they're not just digging in the dirt; they're engaging with a living laboratory where lessons in science, mathematics, and environmental stewardship bloom alongside the plants.

Educational Value

Dry gardens provide an exceptional educational resource, offering tangible lessons in ecological balance and sustainability. In these gardens, students observe firsthand the importance of water conservation—a critical lesson in today's world where water scarcity affects over two billion people globally. These gardens also introduce students to concepts of biodiversity and the roles different plants and insects play in maintaining ecological balance. For many students, this is a compelling introduction to the idea that even small, local actions, like choosing drought-resistant plants or implementing water-saving techniques, can have broader environmental impacts. This kind of education is vital in cultivating a generation that is knowledgeable about and committed to sustainable practices.

Moreover, dry gardens can help in teaching more than just science. They are a base for cross-curricular learning. For instance, math skills are sharpened through tasks like calculating the area for planting, determining the perimeters for irrigation systems, or analyzing growth patterns. Even subjects like art find a place here, as students design garden layouts or study the aesthetics of plant arrangements. The versatility of dry gardens makes them a powerful educational tool that extends learning beyond the traditional classroom walls.

Student Involvement

The involvement of students in the creation and maintenance of dry gardens fosters a profound connection with nature and a sense of responsibility for the environment. Schools that incorporate garden-based learning often report increased student engagement and enthusiasm for school. Projects that allow students to design, plant, and tend dry gardens provide them with a sense of ownership and pride in their accomplishments. These projects also teach responsibility, as students must care for the plants and make decisions about water use and plant health.

For instance, a project could involve students in a middle school science class researching native, drought-resistant plants, and then planning and planting a garden based on their findings. Such involvement not only enhances their learning experience but also instills a sense of environmental stewardship. As these students watch their gardens thrive, they witness the tangible results of their planning and care, reinforcing the impact of their actions on the environment.

Curriculum Integration

Integrating dry gardening into school curriculums can transform theoretical environmental education into practical, impactful learning. For example, biology classes can use the garden to study plant physiology and adaptation strategies of drought-resistant plants, while environmental science classes might analyze the garden's ecosystem and its role in local biodiversity. Teachers can use the garden as a context for lessons on climate change, sustainability, and human impact on the environment, making these complex topics more relatable and understandable.

Integrating dry gardening into the curriculum can also help schools meet educational standards in science and environmental education. By using the garden as a teaching tool, educators can provide students with experiential learning opportunities that are often more engaging and memorable than traditional classroom activities. This approach enriches the students' educational experience and promotes lifelong learning and curiosity about the natural world.

Success Stories

Several schools have already seen remarkable benefits from integrating dry gardens into their educational programs. For instance, a primary school in New Mexico transformed an unused plot of land into a thriving dry garden with the help of its students. The garden has become an integral part of the school's STEM curriculum (Science, Technology, Engineering, and Math) with students regularly participating in experiments that explore soil science, water conservation, and plant biology. The success of this garden has not only enhanced students' learning experiences but also brought the community together, with local experts frequently invited to speak about sustainable gardening and environmental conservation.

Another success story comes from a high school in California, where students designed and implemented a dry garden that serves as both a scientific study area and a peaceful retreat for students and staff. The garden features a variety of drought-resistant plants and a solar-powered irrigation system, which students help maintain. This project has not only taught students valuable skills in horticulture and environmental stewardship but has also earned the school recognition for its commitment to sustainability.

These examples demonstrate the powerful role that dry gardens can play in education, providing students with practical skills and knowledge that prepare them for the challenges of a changing world. They show that when schools invest in such initiatives, the benefits extend beyond the classroom, enhancing student engagement, supporting community involvement, and promoting a sustainable future.

As this chapter closes, reflect on the transformative role of dry gardens in educational settings, not just as tools for teaching about the environment but as platforms for broader life lessons in responsibility, sustainability, community engagement, and feelings of satisfaction and well-being. These gardens are more than just patches of land; they are nurturing grounds for growing minds, cultivating not only plants but also the seeds of environmental stewardship in young hearts and minds. As we turn our attention to the next chapter, let's carry forward the understanding that every small step in our gardens can lead to significant leaps in our collective environmental consciousness.

Chapter 10

Building a Community of Dry Gardeners

Imagine walking through a neighborhood where every garden is a vibrant testament to sustainability, each plot brimming with drought-resistant plants thriving under the careful watch of passionate gardeners. This vision is not just a dream but a potential reality that begins with fostering a community of like-minded enthusiasts who are eager to embrace and promote dry gardening practices. The creation of such a community not only enriches individual gardening experiences but also cultivates a collective wisdom and resource pool that can transform entire neighborhoods and beyond.

10.1 Starting Local Dry Gardening Groups and Workshops

Initiating Local Interest

The genesis of any community group lies in sparking interest and gathering like-minded individuals. Starting a local dry gardening group begins with outreach. You might consider writing articles for local newspapers or online platforms, posting on community bulletin boards, leaving notices at local grocery and hardware stores, as well as gardening centers and nurseries, and using social media to share the benefits and beauty of dry gardening. Hosting a free introductory workshop at a local library or community center can also attract people of all gardening levels.

It's about planting a seed of interest in sustainable gardening practices. Once people see what's possible, their enthusiasm will grow.

To gauge interest effectively, surveys can be a useful tool. Simple questionnaires distributed online or at local events can help you understand what aspects of dry gardening excite your neighbors the most and what skills they might want to learn. This information is invaluable in tailoring your group's activities to the needs and curiosities of its members, ensuring lively participation and sustained interest.

Organizing Workshops

Workshops are the educational backbone of any gardening group. They provide a platform for sharing knowledge, discussing challenges, and brainstorming solutions. When organizing workshops, it's essential to cater to a range of expertise levels, from beginners to seasoned gardeners, ensuring that content is relevant and engaging for all participants. Topics might include water-efficient irrigation techniques, choosing the right plants for the local climate, or innovative ways to improve soil health.

Collaboration with local nurseries or gardening centers can provide both a venue and an expert speaker for your workshops. Such businesses are often eager to support educational initiatives that promote gardening, and they can offer practical insights and materials. Additionally, hands-on workshops where participants can directly engage with the gardening practices being discussed—such as building a rain garden model or setting up a drip irrigation system—can be particularly effective. These interactive elements not only make learning more engaging but also empower attendees to apply their new knowledge in their own gardens.

Collaboration With Local Experts

No one is an island, and the success of a gardening group often hinges on the strength of its collaborations. Reach out to local horticulturists, environmental scientists, and experienced dry gardeners to share their knowledge and experiences. These experts can provide deeper insights into the science behind sustainable practices and offer practical advice drawn from their professional and personal experiences with dry gardening.

Organizing regular guest lectures or Q&A sessions with these experts can enhance the credibility of your group and provide members with access to advice that is both scientifically sound and locally applicable. This collaboration not only enriches the group's knowledge base but also strengthens community ties to local businesses and educational institutions involved in environmental stewardship and sustainability.

Building a Supportive Network

The heart of any community group is its network—people who share resources, knowledge, and encouragement. Facilitating an environment where members feel comfortable sharing their successes, challenges, and resources is crucial. Encourage members to bring seeds, cuttings, or excess materials to share with the group, fostering a culture of generosity and mutual support.

Creating a resource library, whether physical or digital, can also be incredibly beneficial. This library could include gardening books, design tools, and guides on dry gardening practices that members can borrow. Additionally, setting up a digital forum or email group where members can ask questions, share updates on their garden progress, and post relevant articles or resources helps maintain engagement and ensures that help is always just a message away.

Reflective Exercise

Consider what plants or resources you could contribute to a community gardening group. Reflect on how sharing these could help not only improve your gardening practices but also support and inspire others in your community.

10.2 Leveraging Social Media to Share and Learn

In the digital age, the power of social media as a tool for learning, sharing, and connecting cannot be underestimated, particularly in the niche realms of hobbies like dry gardening. Creating an online platform dedicated to this purpose can transform the way you and your peers explore and expand your gardening practices. Social media groups or pages serve as vibrant community hubs where enthusiasts from around the corner or across the globe can converge to exchange ideas, experiences, and inspiration. There are already many groups devoted to a plethora of different areas of gardening. If you don't find one that meets your specific preferences, you can always form one of your own.

Setting up a social media group dedicated to dry gardening involves several thoughtful steps. First, choose the right platform—Facebook, for instance, offers easy group setup options, extensive reach, and tools such as polls, photo albums, and event pages that are conducive to group interaction. When setting up your group, make it a point to create clear, engaging, and informative descriptions that outline the group's intentions and guidelines. This clarity will help attract members who are genuinely interested in dry gardening. Additionally, regular updates, prompt responses to queries, and scheduled posts about different dry gardening topics can keep the group active and informative.

Encouraging members to share their success stories forms the core of the group's activity. When members post before-and-after photos of their garden transformations or share innovative gardening solutions they've devised, it does more than just showcase their achievements—it inspires others. These real-life examples serve as tangible proof of what can be accomplished with the right techniques and a bit of creativity. To facilitate this, you could organize monthly challenges or themed photo-sharing weeks like "Water-Saving Wednesdays," where members can post pictures of their irrigation solutions or drought-resistant plant setups.

The interactive aspect of social media makes it an excellent venue for hosting online Q&A sessions with experts in dry gardening. These sessions can be scheduled monthly or quarterly and can cover a range of topics from soil health in arid climates to the best drought-resistant plant species for different regions. Utilizing live video features or dedicated post threads for these Q&A sessions can facilitate real-time interaction, allowing members to receive immediate responses to their queries. This direct access to experts not only enriches the learning experience but also helps in building a rapport among community members and experts, fostering a learning environment that is both supportive and enriching.

Another dynamic way of using social media to enhance community learning is by organizing virtual garden tours. Members can share videos or live streams of their gardens, explaining the techniques they use, the plants they grow, and the challenges they face. These virtual tours can offer an inside look at the practical implementation of dry gardening techniques in diverse settings, providing members with ideas and insights that they can adapt to their own garden spaces. To make these tours more interactive, other members can ask questions in real-time or offer suggestions, creating a collaborative and engaging learning experience.

By leveraging social media effectively, you not only expand your knowledge and network but also contribute to a growing community of eco-conscious gardeners. This digital platform offers a unique space where learning is continuous and collective, and where every member can play a role in shaping a more sustainable and beautiful world, one garden at a time.

10.3 Organizing Dry Gardening Events and Garden Tours

Planning and executing community events focused on dry gardening can serve as a catalyst for wider acceptance and excitement around sustainable gardening practices. These events, including plant swaps, seed exchanges, and educational seminars, provide unique opportunities for hands-on learning and community building. To start, consider a plant swap event, which not only encourages community interaction but also promotes the genetic diversity of plant species well-suited for dry environments. Begin by selecting a date and a public venue like a community center or park that supports easy access for all community members. Coordination with local gardening clubs can amplify attendance and participation, leveraging their networks to ensure a successful event. It's essential to communicate clearly the kind of plants or seeds suitable for swapping, focusing on drought-resistant varieties that will thrive in local climates.

When it comes to seed exchanges, these can be particularly enlightening as they often introduce gardeners to new plant varieties. This event could be structured around a mini-seminar where experts can speak briefly about the benefits and care instructions for the seeds being exchanged. Partnering with local agricultural extensions or botanical gardens can provide speakers who bring a wealth of knowledge and enhance the educational value of the event. Before the event, gather seeds from local nurseries or community gardens, and package them in small, labeled envelopes. This preparation allows participants to know exactly what they're getting and how best to nurture the plants from seed to maturity.

Beyond swaps and exchanges, organizing educational seminars can delve deeper into specific aspects of dry gardening, such as soil health, water-efficient irrigation techniques, or landscape design using native plants. These seminars offer a platform for experts to share knowledge and for community members to engage with more complex topics that might not be covered in regular garden club meetings. Planning such a seminar involves identifying topics of high interest or particular relevance to your area, sourcing knowledgeable speakers, and creating a schedule that allows for both presentations and interactive Q&A sessions, fostering a dynamic learning environment.

Garden Tour Logistics

Garden tours are a wonderful way to visually and physically demonstrate the principles of dry gardening. They offer real-life examples of what can be achieved in various settings, from residential gardens to larger community green spaces. Start by identifying gardens in your community that exemplify successful dry gardening techniques. Reach out to homeowners, schools, or businesses that have implemented such practices and invite them to participate in the tour. It's important to select a diverse range of gardens that showcase a variety of landscaping styles and solutions to common gardening challenges, such as soil type variations and water management.

Once the participating gardens are confirmed, plan the logistics of the tour. Create a route that efficiently sequences the garden visits, considering factors like distance, walking accessibility, and the time needed at each site. Providing maps and detailed schedules to participants helps ensure a smooth flow throughout the day. Additionally, arranging for transportation, such as carpooling or a rented shuttle, can make the tour more accessible to a broader audience, including those who might not have personal transport or those with mobility issues.

During the garden tours, encourage the garden owners to discuss their strategies and experiences with dry gardening. This direct interaction not only personalizes the experience but also allows for the exchange of tips and ideas that might not be covered in a standard workshop or seminar. Consider providing participants with a feedback form that can help gauge the effectiveness of the tour and gather suggestions for future events. This feedback is invaluable for refining the focus and execution of subsequent garden tours.

Engagement With the Wider Community

While gardeners are naturally a key audience for dry gardening events, extending the invitation to the entire community can have significant benefits. Engaging schools, local businesses, and even municipal councils can elevate the impact of your events. Schools can integrate these events into their science curriculum, providing students with practical learning experiences about sustainable practices. Local businesses, particularly those related to garden supplies and home improvement, might be interested in participating as sponsors or contributors, providing materials or funding in exchange for promotional opportunities.

Additionally, consider hosting events in collaboration with city or town events that focus on environmental sustainability such as sponsoring a booth at a local Home and Garden Show. Aligning your dry gardening event with broader environmental initiatives can draw in a more diverse crowd and connect your efforts with larger city-wide or regional goals. This broader engagement not only fosters a greater appreciation and understanding of dry gardening practices but also embeds these practices into the fabric of the community, paving the way for more sustainable local policies and initiatives.

Fundraising Opportunities

Finally, these events and tours can serve as excellent opportunities for raising funds for community projects related to dry gardening and sustainability. Consider incorporating a small entry fee for tours or selling themed merchandise such as garden tools, books, or locally produced garden products. Another effective fundraising tactic is to organize raffles during your events, with prizes donated by local businesses. Funds raised can be used to support community garden projects, subsidize water-saving devices for low-income households, or sponsor educational workshops and seminars. This not only helps cover the costs of organizing these events but also contributes to sustaining and expanding the community's dry gardening initiatives.

By thoughtfully planning and executing these events and tours, you not only spread the word about the benefits of dry gardening but also strengthen the bonds within your community, creating a collective force that moves towards sustainability and resilience in gardening practices.

10.4 Contributing to Global Sustainability Through Dry Gardening

As we cultivate our local gardens and foster community networks, it's vital to remember that our efforts also contribute to a global movement toward sustainability. Dry gardening, though deeply local in its practice, holds the potential to influence global environmental health. By extending the reach of our local dry gardening initiatives, we can play a part in a worldwide effort to promote ecological balance and resource conservation. This global perspective not only amplifies the impact of our work but also connects us with a broader community of like-minded individuals and organizations dedicated to nurturing the planet.

Educational Outreach and Global Partnerships

One of the most effective ways to expand the influence of dry gardening is through educational outreach that transcends local boundaries. Partnering with international organizations that focus on sustainability can help local groups tap into a network of resources and knowledge that spans the globe. These partnerships might involve collaborating on global campaigns, participating in international conferences, or contributing to educational materials that advocate for sustainable gardening practices worldwide. Such collaborations can also provide opportunities to learn from the successes and challenges of dry gardening practices in different environmental contexts, from the deserts of Australia to the Mediterranean landscapes of Europe.

Imagine, for instance, a series of webinars co-hosted with an international environmental NGO that explores the role of sustainable gardening in climate change mitigation. These sessions could highlight innovative practices from various parts of the world, providing local gardeners with a broader perspective on how small-scale actions contribute to global environmental goals. Additionally, these partnerships can facilitate cultural exchanges that enrich our understanding and appreciation of the global diversity of gardening traditions, all united by a common commitment to sustainability.

Sharing Knowledge on a Global Platform

In today's digital age, the potential to share knowledge across borders is limitless. Creating a digital repository of dry gardening resources that can be accessed from anywhere in the world is one way to contribute to a global knowledge base. This could include instructional videos, case studies, research papers, and design templates that help individuals and communities implement dry gardening techniques in diverse climates and settings.

Moreover, hosting virtual conferences or creating online courses can reach an international audience, providing valuable education on dry gardening techniques and their environmental benefits. These platforms allow for interactive learning and the exchange of ideas, making knowledge about dry gardening accessible to a global audience. Imagine a gardener in Brazil learning about water conservation from a webinar you hosted, or a community in India adapting a soil amendment technique from a case study your group published. These connections not only spread sustainable practices but also foster a sense of global solidarity in the face of environmental challenges.

Collaborative Projects and Research

Engaging in collaborative projects and research with international entities can significantly advance the science and practice of dry gardening. Joint research projects can investigate topics like the efficacy of native plants in conserving water, or the impact of soil health on biodiversity in dry gardens. By pooling resources and expertise, such collaborations can lead to breakthroughs that benefit not just local communities but also contribute to global environmental science.

Additionally, collaborative projects might involve restoring degraded lands using dry gardening principles or developing new, sustainable agricultural practices that can be implemented in arid and semi-arid regions around the world. Such projects not only provide immediate environmental benefits but also serve as practical models for ecological restoration and sustainable development.

Advocacy and Policy Influence

Finally, advocacy is a powerful tool in scaling the impact of dry gardening from local communities to global audiences. By engaging in policy discussions and advocating for legislation that supports sustainable gardening practices, local groups can influence environmental policy at regional, national, and international levels. This might involve campaigning for water conservation laws, promoting policies that support the use of native plants in public landscaping, or advocating for research funding into sustainable agricultural practices.

Participating in policy advocacy ensures that the principles of dry gardening are integrated into broader environmental strategies and legislative frameworks. This can lead to widespread adoption of sustainable practices that not only preserve precious water resources but also enhance the resilience of ecosystems worldwide.

In bringing our local initiatives to a global stage, we not only magnify the impact of our efforts but also join a community of global citizens committed to nurturing the earth. As we conclude this chapter, let us carry forward the message that every garden, no matter how small, is a part of a global tapestry of sustainability. By extending our reach, sharing our knowledge, collaborating on impactful projects, and advocating for supportive policies, we contribute to a healthier planet and a more sustainable future.

Conclusion

As we conclude this explorative journey through the realms of dry gardening, it's essential to reflect on the transformative power that these practices hold. By integrating the water-saving techniques discussed, such as the ingenious use of DIY drip irrigation systems and ancient, yet effective, ollas, you're not just conserving water—you're participating in a global movement towards sustainability. The adoption of drought-resistant and native plants, as we've seen, not only fortifies your garden against aridity but also bolsters local ecosystems, weaving a stronger, more resilient fabric of biodiversity right in your backyard.

From the various chapters, we've uncovered the critical roles of soil health, the strategic use of mulch, and the cutting-edge advancements in gardening technologies like smart irrigation systems. These elements underscore a future where gardening and technology intersect to create efficient, thriving green spaces in the face of changing climates and urban expansion.

The essence of dry gardening stretches beyond the mere act of planting. It's about creating a sustainable interaction between you and the environment. Regardless of whether you dwell in a sprawling suburban home with ample yard space or a cozy urban apartment with just a balcony for greenery, the principles of dry gardening are crafted to be inclusive and adaptable.

Now, I encourage you to take the initial steps. Start small, perhaps with an olla to water your patio plants or a native shrub that demands little but gives much back to its local habitat. Each action you take sows the seeds for a more sustainable future, not just in your garden but as a part of the larger, global tapestry of environmental conservation.

Engage with the vibrant community of dry gardeners. Whether it's through local groups or active online forums, the shared experiences and insights can dramatically enrich your gardening journey. Each challenge shared and success celebrated knits this community tighter, driving innovation and support that spans across borders and cultures.

Remember, every garden, no matter its size or location, is a stroke in the broader picture of ecological conservation. Your efforts to reduce water use and choose sustainable practices are integral to combating the pressing issues of water scarcity and habitat loss. By transforming your garden, you contribute to a wave of change that nurtures and preserves our planet for future generations.

As you continue to nurture your garden, let it be a place of continual learning and adaptation. Experiment with different layouts, embrace new technologies, and always be open to refining your approach as you grow alongside your garden. The path of a dry gardener is one of discovery, filled with the joys of tending the earth and harvesting not just crops but also a deeper understanding of nature's intricate balances.

In closing, carry forward this spirit of innovation and stewardship with optimism and determination. The techniques and strategies you've learned in this book are more than just gardening methods—they are tools for ecological change. You are now equipped, not only to cultivate a thriving and drought-resistant garden, but also to make a tangible, positive impact on the environment. Together, let's cultivate not just gardens, but a legacy of sustainability and resilience.

You've gained the knowledge you need for your gardening journey. Now, help others find the same by leaving an honest review on Amazon. Your feedback will guide fellow Gardeners in their own pursuits and keep the passion for dry gardening alive.

Simply scan the QR code or click the link below to leave your review:

https://www.amazon.com/review/review-your-purchases/?asin=B0DN7VV6W1

Thank you for helping pass on this knowledge. And thank you for joining me on this journey. May your gardens flourish, and may your spirits soar as high as the sunflowers that dance in the summer breeze. Happy gardening!

Glossary

Understanding the terminology used in dry gardening is essential for both newcomers and seasoned gardeners alike. This glossary not only defines these terms but also reflects the breadth and depth of practices involved in creating a sustainable garden in water-scarce environments.

Aquaponics: This innovative system combines aquaculture (raising fish) and hydroponics (cultivating plants in water) in a symbiotic environment. The waste produced by the fish supplies nutrients for the plants, which in turn purify the water that returns to the fish. This closed-loop system is highly water-efficient and can produce a diverse range of foodstuffs in a compact space, making it ideal for urban gardeners looking to maximize their yield.

Biochar: A powerful amendment for garden soils, biochar is produced by burning organic matter in a low-oxygen environment, a process known as pyrolysis. The resulting product is a porous, carbon-rich char that improves soil health by increasing nutrient retention, enhancing soil structure, and providing a habitat for beneficial microbes. Its stability in soil makes it an excellent long-term solution for carbon sequestration, thereby contributing to climate change mitigation.

Coconut Coir: Sometimes referred to as coir, this is a natural fiber extracted from the husk of coconuts. It is commonly used in products like doormats and brushes. In horticulture, coir is especially valued as an alternative to peat moss in soil amendment and seed starting mixes because it helps improve soil aeration and retains moisture effectively.

Coir is also known for its durability and resistance to bacterial and fungal growth, making it a popular choice for gardening and landscaping applications.

Companion Planting: This method involves strategically placing plants that benefit each other close together. Benefits include pest control, pollination, providing habitat for beneficial creatures, maximizing the use of space, and increasing crop productivity. For instance, planting marigolds alongside tomatoes can help repel certain pests while also beautifying the garden space.

Container Gardening: Particularly useful for those with limited space, container gardening involves growing plants in containers instead of planting them directly in the ground. This method offers flexibility in managing soil quality and moisture, and containers can be moved to take advantage of sunlight and shelter as needed. It's an effective technique for apartment dwellers or those with poor soil conditions.

Drip Irrigation: This efficient watering system delivers water directly to the soil at the base of plants through a network of tubes, drippers, and emitters. It minimizes evaporation and runoff, ensuring that water goes precisely where it's needed most. Drip irrigation is particularly effective in dry gardens where water conservation is a priority.

Dry Gardening: This approach focuses on techniques and plant selections that minimize water use. It often involves drought-resistant plants, efficient irrigation strategies, and soil amendments that enhance water retention. Dry gardening is not only practical in arid climates but is also increasingly relevant in regions where water conservation is becoming essential.

Greywater: This term refers to wastewater from non-toilet sources, such as baths, showers, sinks, and washing machines. It can be reused for garden irrigation, significantly reducing the demand for fresh water. Systems to filter and redistribute greywater can be simple, involving only a diversion from the house to the garden, or more complex, incorporating filters and pumps.

Hardscaping: This aspect of landscaping includes non-living elements such as stone walls, wooden arbors, concrete paths, and metal structures. Hardscaping structures can be both functional and decorative, providing essential framework and contrast within the garden.

Hügelkultur: An innovative practice where garden beds are constructed from decaying wood debris and other compostable biomaterials. As the materials decompose, they release nutrients slowly, improving soil fertility. This method also improves water retention in the soil, reducing the need for irrigation.

Hydroponics: This soil-less method of gardening involves growing plants in a nutrient-rich water solution. It is highly water-efficient and can be used to grow plants in environments where traditional gardening is challenging or impossible.

Hydrozoning: This technique involves grouping plants with similar water needs together, which optimizes irrigation efficiency and conserves water.

Leaf mold: Leaf mold is a form of compost made almost entirely of decomposed leaves. It's created by allowing leaves to break down over time, typically in a compost pile or contained area where they are kept moist and allowed to decompose naturally. This process can take anywhere from six months to a couple of years, depending on the conditions and the types of leaves. The resulting material is a rich, crumbly, dark soil-like substance.

Microclimate: Every garden has its unique microclimates depending on factors such as exposure to sun and wind, elevation, and proximity to reflective surfaces. Understanding your garden's microclimates can help you select the most appropriate plants and water conservation strategies.

Mulching: A critical practice in dry gardening, mulching involves covering the soil with organic materials like leaves, straw, or wood chips, or inorganic materials like stones or plastic. This layer helps retain soil moisture, suppress weeds, and improve soil quality as organic mulches decompose.

Native Plants: These are plants that occur naturally in a particular region or ecosystem. Native plants are adapted to local environmental conditions and require far less water, fertilizer, and pesticides than non-natives.

Neem oil: Neem oil is a natural, organic oil extracted from the seeds of the neem tree found in India. It is an effective natural pesticide and also helps control fungal diseases such as powdery mildew, black spot, and rust.

Ollas: An ancient irrigation method that uses unglazed terracotta pots buried in the garden. Water seeps through the walls of the ollas, providing moisture directly to the plant roots nearby. This method minimizes water loss and is highly efficient for dry gardening.

Perlite: A form of volcanic glass that is mined and then heated to a high temperature until it pops like popcorn. This process expands it to many times its original volume, resulting in a lightweight, white, porous material. Perlite is commonly used in horticulture as a soil amendment because of its ability to improve soil structure by enhancing drainage and aeration. It doesn't decompose or break down over time, making it a durable choice for plant root systems.

Permaculture: A design philosophy that mimics the patterns and relationships found in nature. Permaculture aims to create sustainable and self-sufficient agricultural ecosystems and can be applied at scales ranging from the home garden to commercial agricultural operations. See Chapter 2 for more information.

Rain Garden: These gardens are designed to take advantage of rainfall and storm runoff. They are usually planted with native shrubs, perennials, and grasses which thrive in water and help to soak up water, reducing runoff and improving water quality.

Sustainability: This broad concept involves practices that do not deplete resources or harm ecological systems, supporting long-term ecological balance. In gardening, sustainability involves choosing methods and practices that conserve resources, such as water and soil, and reduce waste and pollution.

Vermicomposting: This method uses worms to compost kitchen scraps and other organic waste into rich soil amendments. The process not only recycles waste but also produces vermicompost, a high-quality compost that enhances soil health and fertility.

Vermiculite: A naturally occurring mineral that is heated until it expands into light particles used to improve soil aeration and moisture retention. Vermiculite is especially useful in potting soil mixes for container gardening.

Water Feature: These can be decorative or functional elements in a garden, ranging from small fountains to large ponds. In dry gardening, water features can help increase local humidity and support local wildlife while adding aesthetic value to the garden.

Xeriscaping: This landscaping method specifically reduces or eliminates the need for supplemental water from irrigation. It involves choosing plants that require less water, mulching, and efficient irrigation practices, making it ideal for dry climates or water-restricted areas.

Understanding and utilizing these terms and concepts can significantly enhance your dry gardening practices, making your garden not only more sustainable but also more rewarding and productive. As we continue to explore these practices, keep in mind that each term not only represents a technique or a concept but is also part of a larger philosophy of sustainable living and gardening.

References

Chapter 1: The Foundations of Dry Gardening

- https://www.permaculturenews.org/2020/12/10/ancient-gardening-techniques-from-south-america/
- https://smartwateronline.com/news/the-future-of-water-conservation-innovations-and-technology
- https://www.ncbi.nlm.nih.gov/pmc/articles/PMC6334070/

Chapter 2: Planning Your Dry Garden

- https://www.epicgardening.com/drought-tolerant-plants/
- https://www.agritechtomorrow.com/article/2024/02/revolutionizing-agriculture-innovative-irrigation-solutions-for-a-sustainable-water-future/15247
- https://vlsci.com/blog/soil-additives-for-water-retention-in-soil/
- https://worldpermacultureassociation.com/permaculture-design-in-drylands-expected-results/

Chapter 3: Water-Saving Irrigation Strategies

- https://www.instructables.com/DIY-Drip-Irrigation-in-the-Home-Garden-with-Recyc/

- https://www.nativeseeds.org/blogs/blog-news/how-to-use-olla-irrigation
- https://www.hgtv.com/outdoors/gardens/planting-and-maintenance/DIY-Olla
- https://extension.oregonstate.edu/catalog/pub/em-9101-harvesting-rainwater-use-garden
- https://www.rhs.org.uk/science/gardening-in-a-changing-world/water-use-in-gardens/using-grey-water

Chapter 4: Selecting and Caring for Drought-Resistant Plants

- https://www.epicgardening.com/drought-tolerant-plants/
- https://www.audubon.org/content/why-native-plants-matter
- https://water.unl.edu/article/lawns-gardens-landscapes/water-conservation-vegetable-garden
- https://www.mysoulfulhome.com/my-soulful-home/drought-tolerant-plants-101-guide

Chapter 5: Adapting Gardening Practices to Dry Climates

- https://www.birdsandblooms.com/gardening/drought-tolerant-gardening/drought-tolerant-garden-tips/
- https://www.ncbi.nlm.nih.gov/pmc/articles/PMC8587147/
- https://landscape-water-conservation.extension.org/microclimates-in-water-wise-landscapes
- https://www.phipps.conservatory.org/green-innovation/at-home/greener-gardening-guide/eco-friendly-pest-management-guide

Chapter 6: Technological Advances in Dry Gardening

- https://halff.com/news-insights/insights/water-utilities-newsletter-issue-11-smart-water-irrigation-systems-and-the-benefits-of-using-nativeadapted-landscaping-2/
- https://www.usda.gov/media/blog/2023/01/17/aquaponics-growing-crops-open-water
- https://www.quora.com/How-do-you-create-your-own-DIY-hydroponic-garden-at-home
- https://www.energy.gov/femp/rainwater-harvesting-systems-technology-review

Chapter 7: DIY Projects for the Dry Garden

- https://www.hgtv.com/outdoors/gardens/planting-and-maintenance/DIY-Olla
- https://www.treeandlandscapecompany.com/featured-articles/rain-gardens-for-dry-climates/
- https://www.finegardening.com/project-guides/container-gardening/drought-tolerant-containers
- https://elemental.green/complete-beginner-guide-to-greywater-systems/

Chapter 8: Overcoming Common Challenges

- https://www.phipps.conservatory.org/green-innovation/at-home/greener-gardening-guide/eco-friendly-pest-management-guide
- https://homesteadandchill.com/garden-irrigation-solutions/
- https://www.realsimple.com/home-organizing/gardening/outdoor/drought-tolerant-plants
- https://extension.unr.edu/publication.aspx?PubID=3344

- https://riseservicesinc.org/news/the-benefits-of-gardening-for-people-with-disabilities/?gad_source=1&gclid=CjwKCAjwgfm3BhBeEiwAFfxrG_InLGIfEglP_SVzRAF4Q8Bl6kDzJNY1k5IDb1J7VUZmIXYwnhSMcRoCW98QAvD_BwE

- https://www.quickscream.com/gardening-tools-for-the-disabled/#google_vignette

Chapter 9: Case Studies: Success Stories in Dry Gardening

- https://www.gardendesign.com/small/balcony-garden.html#google_vignette

- https://www.energy.gov/energysaver/landscaping-water-conservation

- https://www.treehugger.com/gorgeous-gardens-using-much-less-water-than-a-lawn-4859091

- https://www.sciencenews.org/article/dry-farming-agriculture-climate-change

Chapter 10: Building a Community of Dry Gardeners

- https://www.aces.edu/blog/topics/business-community-urban/a-guide-to-starting-a-community-garden/

- https://www.ruthbancroftgarden.org/education/2024-dry-garden-design-certificate-program/

- https://www.meltwater.com/en/blog/copy-these-winning-environmental-social-media-accounts

- https://dirt.asla.org/2015/12/11/new-case-studies-on-sustainable-landscape-design/

Made in the USA
Las Vegas, NV
20 December 2024

3e46e64d-f059-47ea-8ca8-e76098d5b110R01